THE NATURE OF CONSCIOUSNESS

THE NATURE OF
CONSCIOUSNESS

Essays on the Unity of Mind and Matter

RUPERT SPIRA

SAHAJA

newharbinger
publications

SAHAJA PUBLICATIONS

PO Box 887, Oxford OX1 9PR
www.sahajapublications.com

A co-publication with New Harbinger Publications
5674 Shattuck Ave.
Oakland, CA 94609
United States of America

Distributed in Canada by Raincoast Books

Designed by Rob Bowden

Printed in Canada

ISBN 978–1–68403–000–2

Library of Congress Cataloging-in-Publication Data on file with publisher

And as imagination bodies forth
The forms of things unknown, the poet's pen
Turns them to shapes and gives to airy nothing
A local habitation and a name.

WILLIAM SHAKESPEARE

CONTENTS

One of the great mysteries of human existence is so basic that most people never think to ask about it: *Can we ever know who we really are?* Simply posing the question runs into an obstacle if we believe that who we are is a walking package of billions and billions of cells. Cells are little bottles of salt water that process chemicals in totally predictable ways. The same goes for brain cells, and no matter how closely you stare at a CT scan or fMRI of the brain, the hot spots that light up seem a long way from Shakespeare and Mozart. Nobody has convincingly shown how glucose – or blood sugar, which isn't all that different from the sugar in a sugar bowl – suddenly learns to think after it passes through a thin membrane and enters the brain.

Rupert Spira belongs to a completely different branch of investigation, which takes 'Who are we?' as an interior question. Being human isn't about cells and chemical reactions but about exploring the essential nature of ourselves and the world. Following this path, even science reaches non-dual conclusions. The great pioneering physicist Max Planck, who coined the term 'quantum', insisted that 'Mind is the matrix of matter'. He elaborated on the point, speaking to a London reporter in 1931: 'I regard consciousness as fundamental. I regard matter as derivative from consciousness. We cannot get behind consciousness. Everything that we talk about, everything that we regard as existing, postulates consciousness.'

Needless to say, modern science didn't follow Planck's lead – quite the opposite. We are in the midst of a headlong rush to solve everything in life through technology and compiling mountains of data for supercomputers to digest. But the total inability to explain consciousness by building it up from molecules, atoms and subatomic particles is a clear failure of science. To claim that discovering more and more complex particles will eventually lead to the emergence of mind is like saying that if you add enough cards to the deck, they learn to play poker.

In short, one can divide the argument between the 'mind first' position and the 'matter first' position. Far and away, the 'matter first' camp prevails at the present moment, since everyone accepts that the physical world 'out there' exists without question. Spira says, in his typically quiet, patient voice, that 'matter first' and 'mind first' are both shortsighted. Taking the simplest possible fact to be true – that there is only one reality – Spira concludes that there is also only one explanation for reality. In these essays he maintains unwaveringly that the only reality is pure consciousness, and everything else, including mind and matter, is a modulation of that reality. A thought is something consciousness *does* – it is not an entity in its own right; likewise an atom. Nature goes to the same place to produce the smell of a rose and a spiral galaxy.

The beauty of this position, which Spira expresses with eloquent conviction, is that the thorny question 'Can we ever know who we really are?' leads to the answer 'Yes'. To be more precise we could say, 'Yes, but...', because finding out who we really are doesn't come in words, but only as an intimate experience, an awakening. And although that experience confronts us at every moment and invites us in, it cannot be compared to any other experience. It lies outside the physical domain and the mental domain at the same time.

Where would such a place be located? Everywhere and nowhere. How do you get there? The journey doesn't require you to go anywhere but here and now. Those answers, however frustrating, are the truth. There's an ancient backlog of discussion on this paradox of starting anywhere and getting everywhere, sometimes called 'the pathless path'. The time-honoured advice, echoed in every spiritual tradition, has pointed inward. The basic notion is that beneath the restless surface of the mind is a deeper level that is unmoving, silent and at peace. This journey relieves our sense of self of all superimposed limitations and reveals its true reality. Illusions fall away. The ego loses its grip. With the experience of the true nature of the Self, a transformation takes place. The key is to transcend our misguided sense of self, and then the light dawns.

In an ideal world, everyone would obey the Old Testament injunction to 'Be still and know that I am God'. Not that religious terms are necessary: the great Bengali poet Rabindranath Tagore declared:

> Listen, my heart, to the whispering of the world.
> That is how it makes love to you.

In other words, intimate contact with the Self is everywhere, and its allure is the same as love's.

If we cannot hear what the world whispers, there is another way, pointed out by Tagore again:

> I grew tired of the road when it took me here and there.
> I married the road in love when it took me Everywhere.

To begin with, the outward world seems to be infinite and inexhaustible, but if we pursue it far enough we inevitably come to the conclusion that it is consciousness itself that is infinite and inexhaustible. The outward journey wears itself out, and then the inward one beckons.

If you try saying this to a sceptic, you run into the same objection: 'Go stand in traffic. When a bus hits you, you're dead. End of story.' Materialists keep insisting that the physical world comes first and that no amount of tricky mental gymnastics can get around that fact. Even sympathetic listeners and committed seekers cling to materialism – perhaps secretly, perhaps guiltily, but mostly, I think, because the full story has yet to sink in. In his gentle but uncompromising way, Spira insists upon telling the full story and beyond that, making it an immediate personal experience.

The full story isn't new. Its origins lie in India's ancient past, although, history and human confusion being what they are, many other stories arose to overlap and muddle it. Someone with knowledge of Indian spirituality will read a few pages of this book – or even just the titles of the essays – and say, 'Ah, Vedanta. That's what he teaches.' But to say this is merely to paste a label on Spira's approach, which also includes the understanding found in the Tantric traditions of Kashmir Shaivism and Dzogchen Buddhism. The Vedas are the sacred scriptures of India, and Vedanta, translated literally, means the end of the Vedas. In other words, Vedanta is the last word in spiritual knowledge, the place you arrive at after absorbing everything else the scriptures can teach you. Vedanta's promise can be stated in a single maxim: 'Know that one thing, by knowing which, all else is known.'

There's enormous appeal in Vedanta's truth-in-a-nutshell, so why didn't it become a kind of universal spiritual path? Why not skip the bulk of spiritual teaching – not just Indian but from all sources – and follow this golden thread? Rupert Spira is rare and all but unique in doing exactly that. In India, Vedanta has a reputation for being complex and intellectual, a subject to which professors and religionists devote their entire lives. What was meant to be practical advice – the one thing you need to know in order for all knowledge to fall into your lap – somehow became abstract and exhausting in its obscurity.

Vedanta needed to be revived for modern people who want practical results; otherwise, the most beautiful truths would be unreachable. Vedanta, to be blunt, was like opening a can of tuna with a piece of limp spaghetti. Spira has been through all that – although he modestly doesn't lean upon his learning – and come out the other side. He has one thing to say because there is only one thing to know: *It's all consciousness.* Because consciousness is creating everything, here and now, and because its creation is endlessly fascinating, he finds beautiful ways to express one thing, often poetically, always compassionately. With a diamond in his hand, he wants to show us every facet.

Forewords risk the pitfall of sounding fulsome, but in all candour, I've gained deeper understanding listening to Rupert Spira than I have from any other exponent of modern spirituality. Reality is sending us a message we desperately need to hear, and at this moment no messenger surpasses Spira and the transformative words in his essays.

Deepak Chopra
September 2016

ACKNOWLEDGEMENTS

I have spent forty years pondering the nature of experience, and this book is the distilled essence of that exploration. Strangely – and perhaps not so strangely – when I was six years old I said to my mother, 'I think our lives are God's dream'. Almost exactly fifty years later this book makes that intuition explicit in rational terms. So I would first like to thank my mother for nurturing this intuition, and also my father, from whom I inherited the means with which to express it.

To do justice to all those to whom I am indebted would take a book in itself, but in spite of this I will try to condense it into a few paragraphs. The warp of my spiritual enquiry and practice has always been the Vedantic tradition, which I first studied under the guidance of Dr. Francis Roles at Colet House in London, which was like a second home for the first twenty years of my adult life. Dr. Roles had received the traditional Advaita teaching from Shantananda Saraswati, the then Shankaracharya of the north of India, whom I consider to be my first teacher.

However, if the warp of my investigation into the nature of reality was made of one colour, the weft contained many. During these years I learned the Mevlevi Turning, a practice of prayer and movement developed by followers of the thirteenth-century Persian mystic Jalaluddin Rumi. This tradition was preserved at Colet House by my late stepfather, Vilhelm Koren, whose presence in our family, though somewhat distant, was a powerful and subliminal influence in my teenage years that conveyed to me the essence of the Sufi tradition. At that time I also learnt Gurdjieff's Movements and studied the writings of the Russian Philosopher P. D. Ouspensky, which had a profound and initiatory effect on me.

During those years, I regularly attended the last meetings of Jiddu Krishnamurti, whose school at Brockwood Park in Hampshire, UK, was close to my childhood home. On one such occasion I found myself standing next to him in the queue for lunch and, to this day, the quality of our encounter left a deeper impression on me than anything I ever heard or read him say. His fierce and tender passion were both an initiation and an incentive in the early days of my investigation.

Also during those years, the teachings of Ramana Maharshi accompanied me on a daily basis, but it was not until I met my teacher, Francis Lucille, that the non-dual understanding became my lived experience. What it is about the relationship with a friend that has the capacity to transform intellectual understanding into felt experience, I do not know. Suffice it to say that everything before that encounter was a preparation for the ongoing revelation of the non-dual understanding that began to unfold under Francis's guidance and friendship.

Until meeting Francis my approach had been primarily a devotional one. Francis introduced me to Atmananda Krishna Menon's incisive lines of higher reasoning, on the one hand, and to the Tantric tradition of Kashmir Shaivism, which he had learned from his teacher, Jean Klein, on the other. Both these introductions opened up new avenues of exploration and experience. Under Atmananda Krishna Menon's meticulous guidance, I felt really free for the first time to think about truth or reality and was, as a result, relieved of the misunderstanding – common among many traditional and contemporary non-dual approaches, and to which I also subscribed in the early years – that thought is inimical to spiritual understanding. From the Tantric approach I learned to take my understanding into the way I felt the body and perceived the world.

However, of the many things I learned with Francis, and for which I am eternally grateful, perhaps the most significant was the realisation that my intense desire to know the nature of reality and my love of beauty were the same endeavour, thus reconciling in me the truth seeker and the artist. In the years I spent with him, I came to understand John Keats's words:

> Beauty is truth, truth beauty, – that is all
> Ye know on earth, and all ye need to know.*

There have been many other influences, too many to be named here, except for Michael Cardew, with whom I apprenticed as a ceramic artist in my

*From 'Ode on a Grecian Urn' (1820).

late teens. No account of my influences would be complete without reference to him. He taught me, without my realising it, the language of form and the process that an artist must undergo, both within himself and in relationship with his materials, if he wants his work to transmit meaning from the maker to the seer or user through lines of cognition that are not accessible to reason. He taught me what it takes to make an object that has the capacity to indicate viscerally, as Cézanne put it, the taste of nature's eternity.

I would like to thank Deepak Chopra for his characteristically generous comments in the Foreword to this book and for his unreserved support of my work. Likewise, Bernardo Kastrup for his penetrating and insightful Afterword, and for the fearless humility with which he extends the subject matter of this book to a field into which I cannot venture. I am also grateful to Mark Dyczkowski, Paul Mills and Peter Fenwick for their kind and generous endorsements.

I would also like to thank all those who, in a more direct way, have been instrumental in the fruition of this book, especially Jacqueline Boyle and Rob Bowden for their endless patience and scrupulous attention to detail, and to Linda Arzouni and Caroline Culme-Seymour for their helpful comments about the manuscript.

Finally, I would like to thank my companion, Ellen Emmet. I am not often lost for words and, as you are about to discover, have yet to transpose into my writing the art of 'less is more' that, at least to some extent, I mastered in my studio. However, when it comes to acknowledging my gratitude to Ellen, I am simply lost for words.

Rupert Spira
October 2016

The Nature of Consciousness

THE HARD PROBLEM OF CONSCIOUSNESS

Our world culture is founded upon the assumption that reality consists of two essential ingredients: mind and matter. In this duality, matter is considered the primary element, giving rise to the prevailing materialistic paradigm in which it is believed that mind, or consciousness – the knowing element of mind – is derived from matter.

How consciousness is supposedly derived from matter – a question known as the 'hard problem of consciousness' – remains a mystery, and is indeed one of the most vexing questions in science and philosophy today. Strangely, the fact that there is no evidence for this phenomenon is not deemed significant enough to dissuade most scientists and philosophers from their conviction that consciousness is a derivative of matter, although more and more are beginning to question it. Most still believe that, with advances in neurology, the neural correlates of consciousness and the means by which it is derived from the brain will sooner or later be discovered, and this belief is reinforced by the mainstream media.

However, until such time, the hard problem of consciousness remains an uncomfortable dilemma for exponents of the materialist paradigm. Ironically, in all other fields of scientific research such lack of evidence would undermine the premise upon which the theory stands, but in a leap of faith that betrays the irrational nature of materialism itself, the conviction at its heart is not undermined by the lack of supporting evidence, nor indeed by compelling evidence to the contrary. In this respect, the prevailing materialistic paradigm shares many of the characteristics of religion: it is founded upon an intuition that there is a single, universal and fundamental reality, but it allows belief rather than experience to guide the exploration and, therefore, the implications of that intuition.

Some contemporary philosophers go further than believing consciousness to be an epiphenomenon, or secondary function, of the brain. In an extraordinary and convoluted act of reasoning they deny the very existence of consciousness, claiming it to be an illusion created by chemical activity in the brain. In doing so, they deny the primary and most substantial element of experience – consciousness itself – and assert the existence of a substance – matter – which has never been found.

In fact, it is not possible to find this substance on the terms in which it is conceived, because our knowledge of matter, and indeed all knowledge and experience, is itself an appearance within consciousness, the very medium whose existence these philosophers deny. Such an argument is tantamount to believing that an email creates the screen upon which it appears or, even worse, that the email exists in its own right, independent of the screen, whose very existence is denied.

*   *   *

For many people the debate as to the ultimate reality of the universe is an academic one, far removed from the concerns and demands of everyday life. After all, reality is whatever it is independent of our models of it. However, I hope that *The Nature of Consciousness* will show clearly that the materialist paradigm is a philosophy of despair and conflict and, as such, the root cause of the unhappiness felt by individuals and the hostilities between communities and nations. Far from being abstract and philosophical, its implications touch each one of us directly and intimately, for almost everything we think, feel and do is profoundly and, for the most part, subliminally influenced by the prevailing paradigm in which we have been raised and now live.

As long as we continue to seek the source of happiness on the part of individuals, and peace amongst communities and nations, from within the existing materialist framework, the very best for which we can hope is to find brief moments of respite from the general trend of experience that is growing ever more divisive. However, there have been epochal moments in history when the collective intelligence of humanity could no longer be contained within the parameters that had evolved over the previous centuries for the purposes of advancing it. The cultural forms that evolve precisely to develop, refine and express humanity's growing intelligence are, at some point, no longer able to accommodate it and become the very means by which it is stifled. The beliefs in a flat earth and a geocentric universe are two such examples.

The idea of a flat earth that prevailed in the ancient world was first challenged by Pythagoras in the sixth century BCE, but it took another two thousand years for his spherical-earth model to be fully accepted by all cultures. Likewise, the idea of a heliocentric universe was first suggested as early as the third century BCE, but it was nearly two thousand years before the Copernican Revolution would make it mainstream.

In each case, a belief that had served humanity's evolution thus far subsequently became the very means of its constraint. But not without resistance! In each case the prevailing paradigm was so tightly interwoven into the ways people thought, felt, acted, perceived and related with one another, and so deeply inculcated into the fabric and mechanism of society itself, that it took two millennia, more or less, for the last vestiges of these ideas to be erased.

In *The Nature of Consciousness* it is suggested that the matter model has outlived its function and is now destroying the very values that it once sought to promote. I believe that the materialist paradigm, which has served humanity in ways that do not need to be enumerated here, can no longer accommodate its evolving intelligence. All around, within ourselves and our world culture, we see evidence that the shell of materialism has cracked. The growing organism of humanity can no longer be accommodated within its confines, and humanity's struggle to emerge is expressing itself in all aspects of society. Nor can its host, the earth, any longer survive its degradation and exploitation.

However, it is no longer sufficient to tinker with the existing paradigm from within its parameters. A new paradigm is required to definitively address the despair and sorrow felt by individuals, the conflicts between communities and nations, and humanity's relationship with nature.

*   *   *

Most revolutions seek to modify the existing state of affairs to a greater or lesser degree but leave the fundamental paradigm upon which they are predicated intact. In *The Nature of Consciousness* another kind of revolution is suggested, one that strikes at the basic assumption upon which our knowledge of ourselves, others and the world is based. It is the revolution to which the painter Paul Cézanne referred when he said, 'The day is coming when a single carrot, freshly observed, will trigger a revolution.'* It is the revolution to which Max Planck, developer of quantum theory, referred when he said, 'I regard consciousness as fundamental. I regard matter as derivative from consciousness.'†

It is the revolution to which James Jeans referred when he said, 'I incline to the idealistic theory that consciousness is fundamental, and that the material universe is derivative from consciousness, not consciousness from the material universe…. In general, the universe seems to me to be nearer to a great thought than to a great machine. It may well be…that each individual consciousness ought to be compared to a brain-cell in a universal mind.'‡ It is the revolution to which Carl Jung referred when he said, 'It is not only possible but fairly probable, even, that psyche and matter are two different aspects of one and the same thing.'§

This revolution is an inner one and addresses the very core of our knowledge of ourselves, upon which all subsequent knowledge and understanding must be based. This book does not explore the implications of this revolution in anything but the broadest terms, but its ramifications touch every aspect of our lives. It is my experience that the implications of the 'consciousness-only' model that is suggested in this book continue to reveal themselves long after the initial insight or recognition itself, gradually colonising and reconditioning the way we think and feel, and subsequently informing and transforming our activities and relationships. It is for each of us to realise and live these implications.

The consciousness-only model is not new. All human beings are at the deepest level essentially the same, therefore there must be a fundamental knowledge of ourselves that transcends the local, temporal conditioning that we acquire from our cultures and thus share with all humanity, irrespective of our political, religious or ideological persuasions. Aldous Huxley referred to this as the 'perennial philosophy', that is, the philosophy that remains the same at all times, in all places, under all circumstances and for all people.

---

* Joachim Gasquet, *Cézanne: A Memoir with Conversations* (1991).
† From an interview published in *The Observer*.
‡ From an interview published in *The Observer*.
§ Jung, C. G., 'On the Nature of the Psyche', in H. Read et al., eds., *The Collected Works of C. G. Jung*, Princeton University Press (1985; original work published 1947).

In the East, the Sanskrit term *sanatana dharma* refers to the same essential, eternal truths that transcend all culturally bound beliefs and customs. *Sanatana dharma*, the perennial philosophy, has been available since the dawn of humanity and has appeared in many different forms and cultures throughout the ages, each culture lending its own particular characteristics to it but never fundamentally changing its original understanding or its essential message for humanity.

Nevertheless, in acquiring the local, temporal conditioning of the cultures in which it appeared, the perennial understanding not only acquired new forms, which is a necessary and inevitable outcome of the transmission of knowledge. It was also inadvertently mixed with ideas and beliefs that belonged to the specific cultures in which it arose and was, as such, modified and diluted to a greater or lesser extent. Even in those cultures in which its essential meaning was not modified or diluted, it was often not fully understood and, as a result, was wrapped in a shroud of mystery which, whilst superficially bearing the hallmarks of wisdom, concealed and sanctified this misunderstanding.

The Nature of Consciousness is also, of course, subject to and a product of the conditioning of the culture and language in which it was written, although the essential understanding that is expressed in it transcends cultural and linguistic conditioning. However, it is my hope that its conditioned form will serve to clarify rather than mystify, obscure or dilute the essential understanding that lies at the heart of the perennial philosophy. I hope in this way to bring the non-dual understanding out of the closet of dogma and esotericism and reformulate it in a way that is accessible to those who seek understanding, peace, fulfilment and friendship beyond boundaries; who do not feel the need to affiliate themselves with any particular group, tradition or religion; and who have become wary of referring to any doctrine, authority or institution at the expense of their own direct experience.

In this book it is suggested that consciousness is the fundamental, underlying reality of the apparent duality of mind and matter, and that the overlooking, forgetting or ignoring of this reality is the root cause of both the existential unhappiness that pervades and motivates most people's lives and the wider conflicts that exist between communities and nations. Conversely, it is suggested that the recognition of the fundamental reality of consciousness is the prerequisite and a necessary and sufficient condition for an individual's quest for lasting happiness and, at the same time, the foundation of world peace.

CHAPTER I

THE NATURE OF MIND

All that is known, or could ever be known, is experience. Struggle as we may with the implications of this statement, we cannot legitimately deny it. Being all that could ever be known, experience itself must be the test of reality. If we do not take experience as the test of reality, belief will be the only alternative. Experience and belief – or 'the way of truth and the way of opinion', as Parmenides expressed it in the fifth century BCE – are the only two possibilities.

All that is known is experience, and all that is known of experience is mind. By the word 'mind' in this context I don't just mean internal thoughts and images, as in common parlance; I mean *all* experience. This includes both our so-called internal experience of thoughts, images, feelings and sensations, and our so-called external experience of consensus reality, that is, the world that we know through the five sense perceptions. Mind thus includes all thinking, imagining, remembering, feeling, sensing, seeing, hearing, touching, tasting and smelling.

If all that could ever be known is experience, and all experience is known in the form of mind, then in order to know the nature or ultimate reality of anything that is known, it is first necessary to know the nature of mind. That is, the first imperative of any mind that wishes to know the nature of reality must be to investigate and know the reality of itself.

Whether mind perceives a world *outside* of itself, as is believed under the prevailing materialist paradigm, or projects the world *within* itself, as is believed in the consciousness-only approach suggested in this book, everything that is known or experienced is known or experienced through the medium of mind. As such, the mind imposes its own limits on everything that it sees or knows, and thus all its knowledge and experience appear as a reflection of its own limitations. It is for this reason that scientists will never discover the reality of the universe until they are willing to explore the nature of their own minds.

Everything the mind knows is a reflection of its own limitations, just as everything appears orange when we are wearing a pair of orange-tinted glasses. Once we are accustomed to the orange glasses, orange becomes the new norm. The orange colour we see seems to be an inherent property of consensus reality and not simply a result of the limitations of the medium through which we perceive. In the same way, the mind's knowledge of anything is only as good as its knowledge of itself. Indeed, the mind's knowledge of things is a *reflection* and an *extension* of its knowledge of itself. Therefore, the highest knowledge a mind can attain is the knowledge of its own nature. All other knowledge is subordinate to and appears in accordance with the mind's knowledge of itself.

In fact, until the mind knows its own essential nature, it cannot be sure that anything it knows or experiences is absolutely true and not simply a reflection of its own limitations. Thus, the knowledge of the ultimate nature of mind through which all knowledge and experience are known must be the foundation of all true knowledge. Therefore, the ultimate question the mind can ask is, 'What is the nature of mind?'

The common name that the mind gives to itself is 'I'. Hence, we say, 'I am reading', 'I am thinking', 'I am seeing', and so on. For this reason, the question 'What is the nature of mind?' could be reformulated as, 'Who or what am I?' The answer to this question is the most profound knowledge that the mind can attain. It is the supreme intelligence.

The question 'What is the ultimate nature of the mind?' or 'Who or what am I?' is a unique question in that it is the only question that does not investigate the *objective content* of the mind but rather the *essential nature* of mind itself. For this reason the answer to this question is also unique. The answer to any question about the objective content of mind will always itself appear as objective knowledge. For example, the question 'What is two plus two?' and the answer 'Four' are both objective contents of mind. But the *nature* of the mind itself never appears in, nor can it be accurately described in the terms of, objective knowledge, just as the screen never appears as an image in a movie.

The mind's recognition of its own essential nature is a different kind of knowledge, a knowledge that is the ultimate quest of all the great religious, spiritual and philosophical traditions and that, although we may not realise it, lies at the heart of each person's longing for peace, fulfilment and love.

*   *   *

Where to begin? As experience is all that could ever be known, we must start with experience, proceeding cautiously, like a scientist, trusting only our observation, doubting every belief and assertion, and only making statements that can be tested and verified by independent observers. If something is true for one person but not another, it cannot be absolutely true. If there is an absolute truth, it must be true for all people, at all times and under all circumstances.

In its search for the absolute truth, science rejects subjective experience on the grounds that it is personal and therefore cannot be validated by anyone other than the person having the experience. For instance, a vision of the Virgin Mary may be true for one person, but many others who have not had the experience will consider it an illusion. However, science has made an error in rejecting all subjective experience on these grounds, for in the ultimate analysis all experience *is* subjective. Therefore, it is not subjective experience but rather *personal, exclusive or idiosyncratic* experience that should be rejected as evidence of absolute reality.

So we could refine the ultimate question as, 'Is there any element of subjective experience that is universal or shared by all?' or 'If the mind only ever knows its own contents, is there any element of the mind's knowledge or experience that is common to all minds?' That knowledge alone would qualify as absolute truth and, therefore, that knowledge alone would serve as the basis of a unified humanity.

Let us agree that there is experience and that experience must be the test of reality. Our experience consists of thoughts, images, memories, ideas, feelings, desires, intuitions, sensations, sights, sounds, tastes, textures, smells, and so on, and each of these is *known*. It is not possible to have a thought, feeling, sensation or perception without knowing it. What sort of experience would be one that is not known? It would not be an experience! Thus, we can say for certain that there is experience and that experience is known, even though we may not know exactly *what* experience is, nor *who* or *what* it is that knows it.

All experience – thoughts, feelings, sensations and perceptions – has objective qualities, that is, qualities that can be observed or measured in some way, have a name and a form, and appear in time or space. It is in this context that I refer to everything in objective experience as 'objects', be those objects apparently physical, such as tables, chairs, trees and fields, or mental, such as thoughts, images, memories and feelings. As such, all objective experience has a form in time or space and, having a form, it has a limit.

But with what is all objective experience known? A thought cannot know a sensation, a sensation cannot feel a perception, a perception cannot see a feeling, a feeling cannot know an image, and an image cannot experience a memory. Thoughts, sensations, perceptions, feelings, images and memories *are known or experienced; they do not know or experience.* Whatever it is that knows objective experience can never itself be known or experienced objectively. It can never be known or observed as an object. It is the *knowing element* in all knowledge, the *experiencing* in all experience. We could say that the mind consists of two elements: its known content and its knowing essence. However, these elements are not actually two separate, discrete entities, and later we will collapse this distinction.

The common name for the knowing or experiencing essence of mind is 'I'. 'I' is the name we give to whatever it is that knows or is aware of all knowledge and experience. That is, 'I' is the name that the mind gives to itself in order to indicate its essential, knowing essence in the midst of all its changing knowledge and experience. I am that which knows or is aware of all experience, but I am not myself *an* experience. I am aware of thoughts but am not myself a thought; I am aware of feelings and sensations but am not myself a feeling or sensation; I am aware of perceptions but am not myself a perception. Whatever the content of experience, I know or am aware of it. Thus, knowing or being aware is the essential element in all knowledge, the common factor in all experience.

'I' refers to the knowing or aware element that remains present throughout all knowledge and experience, irrespective of the content of the known or experienced. Whatever it is that knows the thought 'Two plus two equals four' is the *same* knowing that knows the thought 'Two plus two equals five'. The two thoughts differ and are, as such, amongst the continually changing objects of experience, but each is known by the same knowing subject, irrespective of the fact that one is true, the other false.

Whatever it is that knows the feeling of depression is the same knowing that knows the feeling of joy. The two feelings are different but are known by the same knowing subject, irrespective of the quality of the feeling. The feelings of depression and joy may alternate, but the knowing with which they are known remains continuously present throughout their changes. Whatever it is that knows the sound of birdsong is the same knowing that knows the sound of traffic. The two perceptions differ, and each comes and goes, but they are known alike by the same unchanging, subjective essence of all changing experience. The name 'I' denotes that knowing essence that is common to all knowledge and experience.

I am pure knowing, independent of the content of the known. I am the *knowing* with which all experience is known. I am the experience of *being aware* or *awareness itself* which knows and underlies all experience. Pure knowing, being aware or awareness itself is the essential ingredient of mind – the ever-present, subjective, knowing essence of mind, independent of its always-changing, objective content of thoughts, feelings, sensations and perceptions. Being aware or awareness itself is the knowing in all that is known, the experiencing in all experience.

*　　*　　*

All minds refer to themselves as 'I'. Our Christian names are the names that our parents give to us, but 'I' is the name that the mind gives to itself. Whatever the mind is experiencing, it knows itself as the 'I' that is experiencing it. Throughout the day the mind says, 'I am thinking', 'I am hungry', 'I am cold', 'I am lonely', 'I am tired', 'I am travelling to work', 'I am forty-five years old', and so on. As such, the mind consist of a continuous flow of changing thoughts, images, sensations and perceptions. However, there is one element of the mind – the feeling of being or the experience of being aware – that runs continuously throughout all changing experience.

If, instead of being interested in the continuous flow of changing thoughts, images, sensations and perceptions, the mind becomes interested in its own essential nature, it will discover that the feeling of being or the experience of being aware is the common factor in *all* experience but does not share the particular qualities, characteristics or limitations of any *particular* experience. All the qualities, characteristics and limitations of experience are temporary and ever-changing colourings or modulations of mind but not its essential, irreducible nature.

In other words, as a first step towards realising the essential, irreducible nature of the mind, we separate out the permanent element of experience from its changing forms. We separate out the experience of being aware from what we are aware of.

'I' is the formless or non-objective presence of pure knowing, being aware or awareness itself, which is temporarily coloured by the qualities of experience but not inherently limited by them. 'I am aware', 'I am aware', 'I am aware' runs ever-present throughout all experience. As such, 'I' is the knowing or aware element that underlies and permeates all experience.

*   *   *

All objective experience changes continually. Thoughts, feelings, sensations and perceptions are in a constant state of flux. A thought is by definition always flowing, a feeling always evolving, a sensation always pulsating and a perception always changing, albeit at times imperceptibly slowly. In fact, later we will see that we never actually experience a discrete object such as *a* thought, feeling, sensation or perception, let alone a mind, body or world. But for the time being let us agree that all experience continually changes.

However, each changing thought, feeling, sensation or perception is registered by the *same* knowing 'I', the common element in all experience. The knowing 'I' that is seeing or knowing these words is the same knowing 'I' that was knowing or aware of whatever 'I' was experiencing an hour ago, last week, last month, last year or ten years ago. That knowing 'I' – consciousness or awareness* itself – is the common ingredient in all experience. It remains the same throughout all experience.

Each of us feels that we have always been the same person, although the experience of the body and mind, which we normally consider to be ourself, is continually changing. All we know or experience of the body are changing sensations and perceptions, and all we know of our mind† is a

---

*The terms 'awareness' and 'consciousness' are used synonymously throughout this book.
† The word 'mind' is used here in the conventional sense, to indicate thoughts, images and feelings.

flow of concepts, images and feelings. In fact, the body never knows itself as 'I'. It is the mind that calls itself 'I'. So when I say, 'We have always been the same person', I mean that the mind recognises that there is something in its own experience of itself that always remains the same. Thus, although everything we have ever identified as ourself has changed innumerable times in our lives, each of us feels that there is some part of ourself that remains consistently present throughout all experience.

When we say 'I' today we refer to the same 'I' that we were two days ago, two months ago, two years ago or twenty years ago. What part of our experience of ourself accounts for the feeling of always being the same person? What is it in our experience of ourself that always remains the same? Only the knowing with which all changing knowledge and experience are known. Only the experience of being aware or awareness itself. Only 'I'.

The known or experienced always changes, but the *knowing* with which all changing experience is known always remains the same. When we were five-year-old girls or boys the experience of our parents, home and garden was *known*. As a ten-year-old child the experience of our friends, teachers and classroom was *known*. As a teenager, our first kiss, our studies and the parties we went to were *known*. As an adult, our activities and relationships are always *known*. The current experience – these words, the thoughts and feelings they provoke, sensations of the body and perceptions of the world – are being *known*. All experience is *known*.

Experience never ceases to change, but 'I', the knowing element in all experience, never itself changes. The knowing with which all experience is known is always the *same* knowing. Its condition or essential nature never changes. It is never modified by what it knows. Being the common, unchanging element in all experience, knowing, being aware or awareness itself does not share the qualities or, therefore, the limitations of any *particular* experience. It is not mixed with the limitations that characterise objective experience. It is, as such, unqualified, unconditioned and unlimited.

The knowing with which a feeling of loneliness or sorrow is known is the same knowing with which the thought of a friend, the sight of a sunset or the taste of ice cream is known. The knowing with which enthusiasm or exuberance is known is the same knowing that knows our darkest feelings and moods. The objective element of experience always changes; the subjective element never changes. The known always changes; knowing never changes.

This knowing 'I' – the experience of simply being aware or awareness itself – is never itself either exuberant or sorrowful. Being the common element in both experiences, it is not qualified, conditioned or limited by either. In both experiences, indeed in all experience, it remains in the same pristine condition, without qualification or limitation. The knowing with which exuberance or sorrow is known is not itself changed, moved, harmed or stained by the exuberance or sorrow itself. When the exuberance or sorrow passes, the same knowing remains present to know or be aware of the next object of experience, be it the thought of a friend, the sight of a sunset or the taste of ice cream.

Nothing ever happens to the knowing with which all experience is known. It is not enhanced or diminished by anything that it knows or experiences. When a feeling of sorrow appears, nothing is added to the knowing with which the sorrow is known. When the sorrow leaves, nothing is taken away from it. If any thought, feeling, sensation or perception were identical to our essential nature of pure knowing, then every time a thought, feeling, sensation or perception disappeared we would feel that a little bit of ourself disappeared with it. Indeed, if thought, sensation or perception were inherent to the essential nature of mind or pure knowing, it would not be possible for a thought, sensation or perception to appear, because what is essential to mind must always and already be present within it and as it. Therefore, the essential nature of mind does not appear or disappear; it has no beginning or end. It was not born and will not die.

We always feel essentially the same whole, indivisible, consistently present person, only we mistake the essential nature of that person. Although innumerable thoughts, feelings, sensations and perceptions are added to us and subsequently removed from us during the course of our lives, the person or self that we essentially are remains always the same. That is, *pure knowing*, the essence of mind, 'I', always remains in the same pristine condition.

Exuberance, enthusiasm, sorrow, loneliness, the thought of a friend, the taste of ice cream, and so on, are not separate from the knowing of them – not separate from 'I' – but neither are they identical to it. The knowing with which all experience is known is to experience as a self-aware screen would be to a movie – that is, a magical screen that is watching the movie that is playing upon it. The movie is not separate from the screen, nor is it identical to it. Our changing thoughts, feelings, sensations and perceptions colour our essential being of pure knowing or awareness itself, but they do not modify, qualify, condition or limit it, nor are they identical to it.

It is for this reason that the essential nature of mind is said to be *pure* knowing or *pure* awareness. 'Pure' in this context means unmixed with any of the qualities, conditions or limitations that it knows or is aware of, just as the screen is not inherently mixed with any of the limited forms that appear in a movie. The essential nature of mind – the experience of being aware, pure knowing or awareness itself – is inherently unconditioned and unlimited.

Likewise, just as a screen is never disturbed by the drama in a movie, so pure knowing, being aware or awareness itself is never disturbed by experience, and thus it is inherently imperturbable or peaceful. The peace that is inherent in us – indeed that *is* us – is not dependent on the content of experience, the circumstances, situations or conditions we find ourselves in. It is a peace that is *prior to* and at the same time *present in* the fluctuations of the mind. As such, it is said to be the peace that 'passeth understanding'.

* * *

Whatever it is that knows, experiences or is aware of all experience is the most intimate, essential and irreducible nature of mind, 'I' or our self. Knowing or being aware is not a quality *of* our self; it *is* our essential self. Our self doesn't *have* or *possess* awareness; it *is* awareness or consciousness itself. The suffix '-ness' means the existence, state, presence or being of, so the words 'awareness' and 'consciousness' imply the presence of that which is aware or conscious.*

The danger of using a noun to denote the experience of being aware or pure knowing is that we reify or objectify something – which is not a thing – that we have already discovered to be without objective quality. Conventional language has evolved to describe objective experience, and in using the terms 'awareness' and 'consciousness' we are borrowing elements of conventional language and adapting them to a purpose for

---

* Being 'conscious' in this context is not meant in the conventional sense of being aware of an external object or a thought or feeling, but rather the simple experience of being aware, independent of objects.

which they were not intended. In fact, if we really want to speak the absolute truth we should remain silent, as indeed some do.

However, others amongst us who feel compelled to articulate reality in words try to make the best use of these ill-adapted symbols, using them as skilfully as possible and in a way that evokes the reality of experience without ever confining it within the limits of language. Others speak the language of poetry, and portray the relationship between the objective elements of experience and the essential nature of mind as a play of separation and union between a lover and her beloved, thereby avoiding having to frame reality within the confines of reason.

All experience is known, and therefore pure knowing, being aware or awareness itself is present in all experience. It would not be possible to have or know experience if knowing or awareness were not present. As such, awareness is the prerequisite for all experience; it is the primary and fundamental element in all experience. We cannot legitimately assert the existence of anything prior to awareness or consciousness, for if such an assertion were based on experience rather than belief, awareness itself would have to be present to know the experience, and therefore that experience would not be prior to it.

In fact, we can go further than this. Not only is pure knowing or awareness itself the *primary* element of mind; it is the *only* substance present in mind. It is easy to check this in experience. All that is or could ever be known is experience, and all there is to experience is the knowing of it – in fact, not the knowing 'of it', because we never encounter an 'it' independent of knowing. All there is to 'it' is the experience of knowing.

In other words, we never know anything other than knowing. All there is to experience is knowing. There is no object that is known and no subject that knows it. There is just knowing. And what is it that knows that there is knowing? Only that which knows can know knowing. Therefore, only knowing knows knowing. That is, awareness or consciousness is all that is ever known or experienced, and it is awareness or consciousness that is knowing or experiencing itself. Thus, the only substance present in experience is awareness. Awareness is not simply the *ultimate* reality of experience; it is the *only* reality of experience. Experience is a freely assumed self-modulation of awareness itself, but whatever the content of the modulation, at no time does any substance other than awareness ever come into existence.

The word 'reality' is derived from the Latin *res*, meaning 'thing', betraying our world culture's belief that reality consists of things made of matter. However, nobody has ever experienced or could experience anything outside awareness, so the idea of an independently existing substance, namely matter, that exists outside awareness is simply a belief to which the vast majority of humanity subscribes. It is the fundamental assumption upon which all psychological suffering and its expression in conflicts between individuals, communities and nations are predicated. If we refer directly to experience – and experience alone must be the test of reality – all that is or could ever be known exists within, is known by and is made of awareness alone.

Any intellectually rigorous and honest model of experience must start with awareness, and indeed never stray from it. To start anywhere else is to start with an assumption. Our world culture is founded upon such an assumption: that matter precedes and gives rise to awareness. This is in direct contradiction to experience itself, from whose perspective awareness is the primary and indeed only ingredient in experience, and must therefore be the origin and context of any model of reality.

CHAPTER 2

ONLY AWARENESS IS AWARE

Our world culture is founded on the assumption that the Big Bang gave rise to matter, which in time evolved into the world, into which the body was born, inside which a brain appeared, out of which awareness at some late stage developed. None of this could ever be verified, because it is not possible to legitimately assert the existence of anything prior to awareness or consciousness. Therefore, any honest model of reality must start with awareness. To start anywhere else is to build a model on the shifting sands of belief.

It is commonly believed that awareness is a property of the body, and as a result we feel that it is 'I, this body' that knows or is aware of the world. That is, we believe and feel that the knowing with which we are aware of our experience is located in and shares the limits and destiny of the body. This is the fundamental assumption of self and other, mind and matter, subject and object that underpins almost all our thoughts and feelings, and is subsequently expressed in our activities and relationships.

However, it is not 'I, *the body*' that is aware; it is 'I, *awareness*' that is aware. A body doesn't have awareness; awareness 'has' the experience of a body. The body, as it is actually experienced, is a series of sensations and percep-tions in the finite mind, and the only substance present in mind is pure knowing or awareness. It is thought alone that conceptualises and, as such, abstracts the body as an object made of matter appearing outside awareness. However, if we stay strictly with the evidence of experience, the body is an appearance in awareness; awareness is not an appearance in the body.

An inevitable corollary to the belief that awareness is a by-product of the body is the belief that awareness is intermittent, that it appears and disappears, that it starts at one time and ends at another. However, to assert the absence of awareness as an actual experience, something would have to be aware of that experience, and that very 'something' would be awareness itself. Therefore, such a claim confirms the *presence* of awareness rather than its absence. It is our experience that we are continuously aware.

When I say that *we* are continuously aware, one might legitimately ask who is the 'we' that is being referred to. Who is the 'we' that is aware that we are continuously aware? Who or what has the experience of being aware? Who or what knows that there is awareness? Awareness is the aware or knowing element in all experience and is, therefore, the only 'one' present to know whatever is known or experienced, including its own presence.

Therefore, the experience of being aware, or the knowledge 'I am', 'I am aware' or 'There is awareness' is *awareness's knowledge of itself.* Only awareness knows that there is awareness. Only awareness is aware. As such, awareness is *self-aware.* Just as all objects on earth are illuminated by the sun but the sun alone is self-luminous, so all experience is known by awareness, but awareness itself is self-knowing. Thus, it is *awareness's* experience that it is continuously or, more accurately, eternally aware.

Being aware is awareness's *primary* experience. Awareness knows its own being before it knows any other thing. Thus, awareness's knowing of its being is the original knowledge, the primary, fundamental and subjective knowledge upon which all objective knowledge is based. It is the ground from which all experience rises and upon which it rests, just as the colourless screen is the foundation upon which all images play.

Awareness's knowing of its being is also its *ultimate or final* knowledge, that is, the knowledge that remains over after every thought, feeling, sensation and perception has vanished, just as the screen remains over after a movie ends. It is to this understanding that Jesus refers in the Book of Revelation when he says, 'I am the Alpha and the Omega, the First and the Last, the Beginning and the End.' It is also the knowledge to which the term Vedanta, meaning the 'end of knowledge', refers.

* * *

Because we normally believe that it is 'I, the body' that is aware or has awareness, the body and, by extension, the world are considered to precede awareness. Thus, awareness is considered to be derived from the body, as an epiphenomenon of the brain. However, in order to legitimately claim this, we would have to experience the body prior to the experience of being aware, and then notice the experience of being aware arising *in the body*. Nobody has ever had, or could ever have, this experience. If we maintain the honesty and rigor of the scientist, who is willing to state only the facts of experience without any regard for their implications or consequences, we must acknowledge that awareness is the primary element in all experience.

Being aware or awareness itself is not a property of a person, self or body. All that is known of a body is a flow of continuously changing sensations and perceptions. All sensations and perceptions appear in mind, and the only substance present in mind is awareness or consciousness itself. Thus, the body is an appearance in mind, and the ultimate reality of mind, and therefore the body, is awareness.

The essential nature of awareness is to be aware, just as it is the nature of the sun to shine. Simply by being itself, awareness is aware of itself, just as the sun illuminates itself simply by being itself. Awareness cannot cease being aware, for being aware is its nature. If it ceased being aware, it would cease being awareness. Where would awareness go if it were to cease and therefore disappear? There is nothing in our experience – that is, there is nothing in awareness's experience – that is prior to or 'further back' than awareness itself, into which awareness could disappear.

It is thought that mistakenly identifies awareness with the limits and destiny of the body and thus believes that awareness is intermittent. However, in awareness's own experience of itself – and awareness is the only 'one' that is in a position to know anything about itself – it is eternal, or ever-present.

Although awareness is eternally aware of itself, it is not always aware of the body. The body is an appearance in and of the finite mind, and the finite mind is itself a modulation of awareness. Thus, the body is a temporary modulation of and an appearance in awareness; awareness is not an appearance in the body. Awareness itself is not intermittent. It is a continuous, or, more accurately, ever-present, non-objective experience.

How could something that is ever-present be a by-product of something that is intermittent? To believe that awareness is a by-product of the body is like believing that a screen is produced by the movie that appears on it. The screen is continuous; the movie comes and goes. The movie is a by-product of the screen. Awareness is like a self-aware screen: in its own experience of itself it is continuous or ever-present.

* * *

Awareness vibrates within itself and assumes the form of the finite mind. The finite mind is therefore not an entity in its own right; it is the *activity* of awareness. There are no real objects, entities or selves, each with its own separate identity, appearing in awareness, just as there are no real characters in a movie. There is only awareness and its activity, just as there is only the screen and its modulation.

Awareness assumes the form of the finite mind by identifying itself with the body, through the agency of which it knows the world, in the same way that at night our own mind collapses into the mind of the dreamed character from whose point of view the dreamed world is known. Just as the activity of our individual mind appears to itself in the form of the dreamed world, so awareness appears to itself as the world in the form of the activity of each of our minds. It is only from the point of view of the apparent awareness-in-the-body entity – the finite mind – that awareness now seems to be limited and temporary, and that the body and world seem to have their own independent status as objects.

Awareness assumes the form of the finite mind in order to simultaneously create and know the world, but it doesn't need to assume the form of mind in order to know itself. Awareness is *made* of pure knowing or being aware, and therefore knows itself simply by being itself. Awareness doesn't need to reflect its knowing off an object in order to know itself, just as the sun doesn't need to reflect its light off the moon in order to illuminate itself.

A child sometimes takes a mirror and catches the light of the sun with it, reflecting the sun's light into a friend's eyes. To believe that awareness needs a finite mind to know itself is like believing that the sun illuminates itself by reflecting its light off a little piece of mirror. The sun doesn't need a mirror to illuminate itself; it illuminates itself by itself. Likewise, awareness doesn't need to shine in or on an object, such as a mind or a body, to know itself through the reflected light of that object. The only substance that is present in awareness is being aware, aware being or pure knowing. Therefore, the knowing of itself is what it *is*, not what it does.

In order to illuminate an object, the sun must direct its rays away from itself, towards that object, but in order to illuminate itself the sun doesn't have to direct its rays anywhere. Likewise, to know an object, other or world, awareness has to rise in the form of mind, which it does by locating itself in a body, from whose point of view it can now direct the light of its knowing towards that object. But in order to know *itself* it doesn't need to direct its knowing in any particular direction. It doesn't have to go anywhere or do anything. For awareness, being itself *is* knowing itself, just as for the sun, being itself *is* illuminating itself.

All objects and selves are known by awareness, but awareness is known by itself alone. Thus, all objects and selves depend upon and are relative to awareness, but awareness is relative to nothing. All knowledge is relative except awareness's knowing of its own being. Awareness's knowledge of itself is thus absolute. In fact, awareness's knowledge of itself is the only absolute knowledge there is, and is as such the foundation and fountain of all relative knowledge.

Just as the sun is too close to itself to turn round and illuminate itself in subject–object relationship, so awareness is too close to itself to stand apart from itself as a separate subject of experience and know itself as an object. Thus, awareness's knowledge of its own being is utterly unique. It is a category of knowledge that transcends all other knowledge and experience. It is sacred knowledge. It is absolute. It remains the same at all times, in all places, and under all conditions and circumstances. It is the only certainty, from which all other knowledge borrows its relative certainty.

Awareness's knowing of its own being is imperturbable, indestructible, inextinguishable, indivisible, immutable, immortal, invulnerable. It cannot be touched, but all knowledge and experienced is touched by it. It is the only knowledge that does not require the division of experience into an apparent duality of subject and object, and so it is said to be non-dual knowledge.

The self that knows is the self that is known, just as the sun that illuminates is the sun that is illuminated. All other knowledge and experience requires the division of experience into an apparent subject from whose perspective an object, other or world may be known. In relation to all objects, awareness can be said to be the ultimate subject of experience, but its knowing of its own being transcends the duality of subject and object.

The belief that awareness needs a mind in order to know itself is a common misunderstanding in the field known as Consciousness Studies, where the disciplines of non-duality and science meet, and particularly in contemporary expressions of the non-dual understanding. Awareness need only assume the form of an apparently separate subject of experience, the finite mind, in order to know a separate object, other or world. To know *itself* it need not assume the form of a separate subject, nor can it know itself as an object.

For awareness, there is no distance between itself and the knowing of itself. It is simultaneously the subject and the object of its own experience. The essential, irreducible nature of awareness is to be and know itself alone. The knowing of its own objectless being is awareness's primary experience. Just by being itself, it knows itself. Awareness is the knowing element in all that is known or experienced and is, therefore, the only 'one' present to know or experience anything, including itself. The ordinary, intimate and familiar experience of simply being aware is *awareness's awareness of itself.*

*　　*　　*

The mind can never know or find awareness, although everything that it knows or finds is made of awareness alone, just as a character in a movie can never know or find the screen although everything that she knows or finds is made only of the screen. The mind that seeks to know or find awareness is like a character in a movie that travels the world in search of the screen. It like a current in the ocean in search of water. The mind is made out of the very stuff for which it is in search, but it can never find that stuff on its own terms, that is, as an objective experience in time and space.

Imagine physical space prior to the appearance of any object within it, just a vast, borderless space. Now imagine adding to this space the quality of being aware or knowing. The space is now a vast *aware* or *knowing* field, without borders and empty of objects. If we were now to remove the space-like quality from this aware or knowing field, we would end up with pure, dimensionless knowing or being aware, that is, we would end up with awareness or consciousness itself.

In fact, it is not possible to imagine something that has no dimensions. Indeed, something that has no dimensions is not a *thing*. Whatever we can think of must have objective qualities and therefore a dimension in time or space. Awareness itself has no dimensions and is thus not a thing or object of any kind, and yet the experience of awareness or being aware is an undeniable, albeit non-objective experience. However, this does not invalidate the attempt to think of awareness. In trying to imagine the very awareness out of which it is made, the mind will bring itself to its own end, and as a result, objectless awareness will shine as it is.

There is actually no such experience as the ending of mind. Indeed, there is no such thing as *the* mind or *a* mind. The only entity present in mind, if it can be called an entity, is awareness or consciousness itself, and awareness or consciousness never ends, nor indeed starts. Mind only *believes* that awareness starts and ends because it identifies awareness with the limited and temporary body. However, in awareness's own experience of itself – and awareness is the only one that knows anything about awareness – awareness is ever-present.

Mind is the *activity* of awareness. Therefore, in the same way that the screen doesn't come to an end when a movie ends but simply loses its temporary colouring, so awareness doesn't end when the mind stops, but simply ceases colouring itself in the form of mind's activity.

Mind is a self-colouring of awareness, just as a movie is a self-colouring of the screen. In the attempt to know the awareness out of which it is made, the mind simply loses its colour and stands revealed as pure, dimensionless, colourless awareness – pure in the sense that it is not mixed with anything other than itself, and dimensionless in that it has no objective qualities extended in time or space. This zero-dimensional awareness is not an abstraction of thought to which no one has access or knowledge. It is the very awareness with which each of us is currently knowing our experience.

In fact, it is not the awareness *with which* we are knowing our experience. This non-dimensional awareness is not a quality *of* our self, nor does it belong *to* our self. It *is* our self – and not even *our* self. It is *the* self, if it can be called a self. There is no 'me' or 'us' to whom awareness belongs. It belongs to itself.

We do not *have* awareness; we *are* awareness. Awareness is not an attribute of the body, just as the screen is not a property of a character in a movie. Nor is awareness *in* the body; rather, it is 'in itself'. Just as the screen does not appear in the space and time that exist for the character in a movie, so awareness does not appear in the space and time that seem to exist for the finite mind.

As a concession to the mind that wants to think about the nature of awareness, it is legitimate and even necessary to add a subtle space-like quality to it to give it apparently objective and thus conceivable, describable qualities. So, to accommodate our desire to think and speak of awareness, let us conceive of it as a vast, borderless, empty, self-aware space, a field or medium whose nature is simply knowing or being aware. In time, thinking about awareness gives way to the contemplation of awareness – its contemplation of itself – of which more will be said in subsequent chapters.

*   *   *

Everything appears to mind in accordance with its understanding of itself. 'As a man is, so he sees. As the eye is formed, such are its powers.'* It is for this reason that science cannot tell us anything about the nature of awareness. What passes for the increasingly popular field of Consciousness Studies is, in almost all cases, a study of brain activity, not a study of consciousness. Only consciousness knows about consciousness. Only awareness is aware of awareness. Science is an activity of the finite mind, that is, an activity of thought and perception, and necessarily superimposes the limitations of mind upon everything it knows or perceives.

Everything that is known by the mind is an expression and reflection of its own limitations. Being temporary and finite itself, the mind believes awareness to be likewise. Most minds, through which objective reality is known, forget their own limitations and project them instead onto whatever they know or perceive. Thus, everything experienced by the mind appears to be temporary in time and/or finite in space. Forgetting that it has projected its own limitations on reality, the mind believes that the time and space it seems to experience are innate qualities of objective reality itself, whereas in fact they are simply reflections of its *own* limitations.

Time and space are, in fact, dimensionless awareness refracted through the prism of the finite mind, that is, refracted through thought and perception. They are the filters through which awareness perceives its own reality in the form of the world. If reality is refracted through the mind of a flea, it will appear in accordance with the limitations of a flea's mind;

---

*William Blake, letter to the Reverend John Trusler (1799).

if through the mind of a dog, in accordance with the limitations of a dog's mind; and if through the mind of a human being, in accordance with the limitations of a human mind.

However, mind is not something separate from reality. It is reality itself – awareness itself – which assumes the forms of each of these minds and through their agency is able to know or perceive a segment of its own infinite potential in the form of the world. In other words, the illusion of a multiplicity and diversity of objects known by a separate subject remains; ignorance of its reality goes. As the eighth-century Zen master Huang Po said, 'People neglect the reality of the illusory world.'

Even when the essential nature of mind has been recognised, reality will still *appear* as a multiplicity and diversity of objects and selves, in accordance with the limitations of the mind through which it is known. However, this appearance will be informed by the understanding that the apparent multiplicity and diversity of reality is not a quality of reality itself, but of the mind through which and as which it is perceived. It will be recognised that the reality that underlies the appearance of multiplicity and diversity is itself an infinite, indivisible whole, and this understanding will inform all the subsequent activities of such a mind.

The mind cannot know the nature of reality until it knows its own nature, thus the science of mind is the highest science. By 'science of mind', I do not mean the study of the *content* of mind; I mean the knowledge of the *essential nature* of mind. The essential nature of mind is that element of mind which remains continuously present throughout all its changing knowledge and experience. It is that element of mind that cannot be removed from it. It is original, unconditioned mind, pure knowing, awareness or consciousness itself. Thus, the ultimate science is the science of consciousness.

However, the science of consciousness is a unique science, because it is the only branch of knowledge that does not require consciousness to rise in the form of the finite mind and, as such, direct itself towards objective knowledge or experience. The science of consciousness is entirely between consciousness and itself. It is about awareness's knowledge of its own being.

Awareness's knowledge of itself is the only absolute knowledge. It is sacred knowledge; in religious language it is God's knowledge of Himself.* It is the highest understanding, upon which all subsequent knowledge must depend.

* Referring to God as 'Him' is used simply as a convention and has no other significance.

CHAPTER 3

PANPSYCHISM AND THE CONSCIOUSNESS-ONLY MODEL

The understanding that only awareness is aware is one of the most challenging aspects of this approach and, at the same time, the most important to grasp.

If we start with the belief that it is 'I, the body' or 'I, the person' that is aware, everything we subsequently know will be conditioned by that belief. I would suggest that the reason contemporary science has so much difficulty fitting consciousness into its model of the universe is precisely because the investigation is founded on the assumption that consciousness is a property of the body.

Having made the assumption that the body or the person is aware, most people in general, and scientists in particular, legitimately assume that animals are also aware. If we tread on our cat's tail it screeches, and from this it is reasonable to conclude that the cat is aware, in this case aware of the pain. The belief that the cat is aware is simply an extension of the belief that the body is aware, or that awareness is an attribute of the body or person.

Reasoning in this way, the scientist who is open to the possibility of fitting consciousness into his model of the universe continues down the animal chain, granting various degrees of consciousness to birds, fish, snails, flies, amoeba, and so on, eventually wondering where to draw the line between what is aware and what is not. Wherever they draw the line poses an uncomfortable question: How do inanimate objects on one side of the line evolve into aware beings on the other? In other words, how does insentient matter give rise to consciousness? This question, known as the hard problem of consciousness, lies at the heart of the debate in science and philosophy today.

The idea that consciousness is derived from inert matter is profoundly inimical to our deepest intuition. Recognising this impossibility, many physicists conclude that a degree of consciousness must be present throughout the universe, and this conclusion leads to the statement, common in the field of Consciousness Studies, that consciousness is fundamental to the universe. This formulation eliminates the need to explain how the universe generates consciousness – that is, it seems to solve the hard problem of consciousness.

The belief that consciousness is fundamental to the universe, which is known in philosophy as panpsychism, does not in fact solve the problem. It simply posits that consciousness is fundamental to *matter*, thereby doing away with the problem of how matter generates consciousness. It doesn't address the relationship between consciousness and matter but merely postpones it. I would suggest that the belief that consciousness is fundamental to the universe is still a subtle form of materialism.

* * *

Panpsychism, the belief that consciousness or mind (in philosophy these two terms are equated, unlike in the non-dual tradition, where they are distinguished) is an essential and fundamental property of things, is not a new idea. It was prevalent in early Greek philosophy. The word 'panpsychism' comes from the Greek *pan-*, meaning 'everything' or 'all', and *psyche*, meaning 'mind' or 'soul'. Aristotle, for example, believed that 'everything is full of gods'.

The belief that all things are full of gods, or that consciousness is fundamental to all things, depends upon the existence of *things*. It starts with a multiplicity and diversity of things! It is equivalent to saying that the screen is fundamental to an image. Although this appears to be a true statement, it contains a misunderstanding, and it is in this subtle misunderstanding that the real problem for contemporary philosophy lies.

To suggest that the screen is a fundamental property of the image is to credit the image with more existence than it deserves. It is to start with the image and work backwards from there to the screen. Likewise, to state that consciousness is a fundamental property of the universe is to start with the universe and work backwards from there to consciousness. In other words, it is to start with the materialist assumption that there is something called a universe.

If we start with the assumption of a universe and try to fit consciousness into that model, we end up with the classic panpsychist statement that all things have a degree of consciousness or, more simply, that the universe is conscious. However, from the perspective of consciousness there is no 'all'. From consciousness's perspective there is just its own seamless, indivisible, unified, infinite whole.

The belief that the universe is conscious is New Age non-duality, and it is this confusion that leads so many people who would otherwise be open to the consciousness-only model to reject it. The belief that the universe is aware is simply an extension of the materialist belief that the body is aware. Fleas are not aware; fish are not aware; dogs are not aware; trees and rocks are not aware; human beings are not aware; the universe is not aware. Only awareness is aware! Only consciousness is conscious.

The word 'universe', from the Latin *uni-*, 'one', and *versus*, 'turned', means 'combined into one; whole'. What is it in our experience of the so-called universe that is whole, one, undivided? Only consciousness! Everything else we know about the universe comprises a multiplicity and diversity of objects. The only element of experience that is one, undivided and whole is consciousness itself, or self-aware being. The universe is not conscious; consciousness *is* the universe!

In fact, the more scientists look for a universe, the less they find it. The more they look for matter, the less like matter it seems to be. Why? Because they are looking for it in objective experience. Sooner or later science will realise that consciousness is the reality for which they are seeking in objective knowledge and experience.

If we want to build a model of reality, we must start with first principles. What is the primary element in all experience? Consciousness! To build a theory based upon anything other than consciousness is to build a house on sand. No matter how well the house may be constructed, it will sooner or later collapse due to the insubstantial nature of its foundation.

The belief that consciousness is fundamental to the universe credits the universe with too much existence. The universe does not *exist!* That is, it does not 'stand out from' consciousness with its own independent reality.* Only consciousness truly *is*. The apparent existence of the universe is consciousness itself – indivisible, self-aware being – refracted

---

*The word 'exist' comes from the Latin *ex-*, meaning 'out of', and *sistere*, meaning 'to stand'.

through the activity of the finite mind. The universe borrows its apparent existence from consciousness, just as the landscape in the movie borrows its apparent reality from the true and only reality of the screen.

*　　*　　*

Matter is the way consciousness appears to itself when viewed through the prism of a finite mind. The finite mind always knows experience in duality, that is, in subject–object relationship, so the object must appear in a way that is distinct from the subject. Without this distinction there could be no manifestation. In other words, manifestation must appear as something *other than consciousness*. In order to distinguish it from consciousness, manifestation must have qualities that consciousness does not have. Consciousness is transparent, empty, non-objective, formless. Therefore, manifestation must seem to be solid, full, objective and with form. This is why the rocks and trees in our dreams seem to be solid.

The reason we believe that the universe exists as an object is that we believe the self exists as a subject. That is, the belief in an external universe is predicated upon our belief that our self, the knowing element in all experience, lives in and is a property of the body. The sand upon which materialism, and by extension panpsychism, is built is our belief in ourselves as temporary, finite minds or entities living in and sharing the destiny and limits of the body.

The scientists and philosophers who subscribe to the materialist assumption that dominates our world culture, as well as those who have moved closer to the consciousness-only model and propose panpsychism as the answer to the hard problem of consciousness, will never find the answer to their questions until they discover the ultimate nature of themselves, that is, until they discover the essential nature of the mind.

Everything that is known by the mind appears in accordance with its own knowledge of itself. As long as we start with the belief that 'I, the body' is aware of experience, we are conducting our investigation on a faulty premise. All our subsequent discoveries will contain this fundamental error more or less subtly concealed within them. Materialism and panpsychism both start with things and proceed from there to consciousness. Both approaches try to graft new understanding onto an old model; they put new wine into old skins. We have to start with the understanding that only awareness is aware. Only consciousness is conscious.

Whilst the panpsychist view may be a welcome and necessary intrusion into the prevailing materialist paradigm, it will, I would suggest, sooner or later have to be abandoned. Of course, new paradigms are not born overnight. It was over a hundred years ago that it was first suggested by Albert Einstein, Max Planck, Niels Bohr, Erwin Schrödinger and others that the observer may affect the observed, opening the debate as to the relationship between consciousness and matter. Although this debate fell into decline during much of the twentieth century, it is enjoying something of a renaissance.

Panpsychism is a stepping stone that will, hopefully, at least usher in a new paradigm in which our model of the universe starts with and is built upon consciousness itself. Sooner or later our culture must wake up from the dream of materialism, of which panpsychism is a subtle extension, and establish consciousness in its rightful place as the absolute reality of all that seems to be. The universe is consciousness itself: one seamless, indivisible, self-aware whole in which there are no parts, objects, entities or selves.

CHAPTER 4

THE INWARD-FACING PATH: THE DISTINCTION
BETWEEN CONSCIOUSNESS AND OBJECTS

In order to know the nature of awareness itself, it is first necessary to distinguish awareness from all that it is aware of. Having established the *presence* of awareness, we will now explore in more detail the process by which awareness itself is distinguished from objective experience, paving the way for the process by which the *nature* of awareness is discovered.

In most people the experience of being aware, or awareness itself, is so thoroughly mixed with the content of objective experience – thoughts, feelings, sensations and perceptions – that it is usually overlooked and, as a result, seems to be missing or at least obscured. How many of us were ever asked by our parents, teachers or professors, 'What is it that is aware of your experience?', 'With what is your experience known?' or 'How is it possible for there to be experience at all?' I have yet to meet anyone who answers 'Yes' to this question. Although everyone is aware, few are aware of the awareness with which experience is known. That is, few are aware of being aware.

To overlook the presence of awareness behind and within all experience is equivalent to overlooking the screen during a movie. Once the screen has been overlooked it is no longer possible to view the movie in the right context. Likewise, once awareness or the original nature of mind has been overlooked, it is no longer possible to know who or what anyone or anything truly is.

All that is or could ever be known is experience, and all there is to experience is the knowing of it. However, the knowing of experience is one phenomenon, not two, so one might legitimately question the validity of separating the knowing element of experience from experience itself, that is, separating awareness from objects. In doing so, are we not falling prey to the very duality that characterises the materialist paradigm we are challenging?

In theory it is not necessary to make this distinction, but our world culture has already conceived a division between mind and matter and attributed ultimate reality to the latter. The distinction between mind and matter is felt at the level of human experience as the separation between ourself and all objects and others, and it is the cause of the existential sense of lack and the fear of disappearance or death that characterise and motivate the separate self or ego. It is in response to this belief, and the pursuit of happiness and security that accompanies it, that it is valid and even necessary to make a distinction between the knower and the known. Later we will collapse this distinction, but it is a useful tool in the early stages of our investigation.

Although experience cannot truly be separated from the knowing or awareness of it, the distinction between awareness and experience is important in this initial investigation in order to establish both the presence and the primacy of awareness – its presence because without awareness there can be no experience, and its primacy because awareness is the essential, irreducible and fundamental reality which precedes, pervades and outlives all objective experience.

Without drawing attention to awareness, the knowing element in all experience, one might reasonably conclude either that the body-mind is the knowing subject of experience, as is conventionally believed, or that only objective experience – thinking, feeling, sensing and perceiving – exists, as indeed many New Age expressions of the non-dual understanding assert. In these teachings it is often claimed that 'There is only this', implying that all that exists is the multiplicity and diversity of objective phenomena, in the absence of a knowing or perceiving subject.

In this view, the fact that it is not possible to have a multiplicity and diversity of objects without a knowing or experiencing subject, just as it is not possible to have only one face of a coin, has been overlooked. In making such a claim, the subject of experience – the ego or separate self – has fooled itself into believing in its own absence. For an ego seeking relief from the discomfort of its own suffering, this belief is a comforting but deluded refuge from which it must at some point return.

*   *   *

How does the distinction between awareness and experience come about? In an infant, experience is an undifferentiated mass in which the knowing element – being aware or awareness itself – is merged so seamlessly with objective experience that there is no sense of being a knowing subject that stands apart from the known object or world in subject–object relationship. In this sense, an infant's experience is similar to that of an animal. A fish has no idea that it is a fish or that it is swimming in an ocean. As far as we know, the fish's experience is an undifferentiated mass of sensing and perceiving.

An infant has no idea that it is an infant, nor that it has, let alone *is*, a body, separate from its mother, lying in a cradle in a room in a world. In the infant's experience of seeing, hearing, tasting, touching and smelling, its knowing of itself, its mother and the world are entirely merged. A multiplicity and diversity of discrete objects known by a separate subject have not yet emerged from the seamless intimacy of pure experiencing. Yet the infant's experience is still known from the perspective of a body, and although the infant is not yet able to conceive of its individual identity, this perspective will later form the basis of its sense of being a separate self.

The pre-verbal, pre-egoic condition of an infant is an early stage of development that New Age thinking and superficial approaches to nonduality often mistakenly equate to the post-egoic recognition that infinite awareness is the fundamental reality of all existence, to which mystics and sages of all cultures refer. It is true that the pre-egoic condition and the post-egoic realisation share qualities of innocence and spontaneity, but we should not confuse the childish state of the former with the childlike condition of the latter. Indeed, young children and animals are almost universally loved because the innocence and spontaneity that we see in them remind us, not of a golden age of childhood which we once knew and have now lost, but of the potential for innocence and spontaneity that lie within us, only thinly veiled or obscured by the clamour of objective experience.

As an infant grows, the sense of itself as a separate, knowing subject of experience begins to emerge and it learns to distinguish itself from all objects and others. To begin with, the self-mother-world matrix is divided into two entities – self-mother and world – and subsequently into self, mother and world, a trinity that will subsequently become the basis of the conventional belief in the soul, God and the world.

The result of this natural separation process is the division of pure experiencing into an experiencer and the experienced, a knower and the known. Self, objects and others begin to arise from within the undifferentiated intimacy of pure experiencing. The infant discovers that the soft, warm sensation that will later be conceptualised as 'my mother's breast' is *not* a part of itself, and that the small patch of waving pink that will later be conceptualised as 'my hand' *is* a part of itself. The seed of the ego or separate self that lies dormant in the infant, and that until now has been a *process* of individuation, begins to crystallise into a discrete *entity*, which identifies itself as the body-mind.

As a perspective, activity or process the ego is neither a mistake nor a problem. However, as an entity it *is* a problem, for the belief that our essential nature is limited to and located in the body is accompanied by the loss of the happiness, freedom and peace that are innate in the knowing of our own essential, irreducible being. That belief is the ultimate cause of the search for happiness through objective experience which defines and dominates the life of the separate self.

The finite mind or ego is, therefore, much more than simply the conceptualisation of the 'I' thought. Even before the infant begins to develop the ability to conceptualise experience, it knows its experience from the limited and located perspective of the body. That is, it knows its experience from the point of view of an apparent self that seems to be located in and to share the limits of the body. This emergence of the ego or separate self is a natural and essential part of development in childhood. It is the process by which the child's identity as a separate individual is progressively established, and if not properly concluded it leads to psychological problems later in life.

As the child grows it begins to conceptualise itself as an entity living in and as the body, and all objects and others are conceptualised in relation to and as an inevitable counterpart of that self. In this way, the *process* of individuation becomes an individual *entity*. Ego as process is replaced by ego as entity.

Under normal circumstances, by the time we are a young teenager we will have established a sense of ourself as an individual body-mind, the separate subject of experience, with a healthy connection to objects, others and our environment. By adulthood, the belief and feeling that the essential, irreducible element of ourself – the experience of being aware or awareness itself – is identical to and thus shares the limits and destiny of the body will have become well established.

In other words, the belief that awareness is both temporary and finite is inculcated into a child at an early stage and becomes the fundamental assumption that underpins and informs its subsequent thoughts, feelings, activities and relationships as it matures into adulthood.

*   *   *

The development of a separate self or ego as an entity is effected by a gradual distinction between the self and all objects and others, a natural process that results in a conventional sense of ourself as a separate individual. In a healthy culture this development of the ego would be seen as an inevitable but transitory stage of development through which an individual passes as it matures. However, in our culture the process of individuation is mistaken for a separate entity, and in most cases all further development is based upon the understanding of ourself as that separate, individuated self or ego.

Although the possibility of transcending the ego is largely unknown in our culture, most people recognise at some level that the ego as a fixed entity represents an unnecessary and unwelcome limitation upon our essential identity and is, as such, the cause of our suffering and its subsequent manifestation in conflicts between individuals, communities and nations. The fact that the term 'egoic' has derogatory implications in common parlance betrays this intuition.

In the Hindu tradition of Advaita Vedanta, the process that distinguishes what we are from what we are not is referred to as *neti neti*, a Sanskrit term that translates as 'not this, not this' or 'neither this nor that'. It is equivalent to the 'via negativa' of Western mysticism in which reality is described in terms of what it is not, rather than what it is. Whilst the development of the infant into a young adult could be considered the initial, instinctive *neti neti* process through which a young person passes during the early stages of life, its further development requires a conscious effort on the part of the adult.

The early stages of the *neti neti* process in childhood progressively distinguish a knowing self from a known object, other or world and culminate in the belief and feeling that we are essentially an amalgam of the body and the mind. We believe and feel that the mind resides in the body, and more specifically in the brain, where thoughts are believed to take place, and that it looks out through the senses to experience the world, which is considered to be separate and independent of itself, made out of something other than itself, namely matter.

This belief is the foundation of our materialist worldview. The materialist paradigm is not based upon the knowledge of matter itself but is a reflection of the fundamental belief in ourself as a temporary, finite mind living in and as the body. The apparent reality of the outside world is an inevitable corollary of the belief in an inside self. As we consider ourself, so we know the world.

The *neti neti* approach, which is initiated naturally in infancy and developed consciously on the spiritual path in adulthood, is the process through which the presence and primacy of awareness are established, and it results in the recognition that what we essentially are is awareness itself. What is, in infancy, the natural, instinctive process by which the developing child distinguishes its own body-mind from its mother and the world is, in adulthood, the same process by which awareness is distinguished from the body-mind.

Awareness is, in fact, never, from its own point of view, mixed with the limitations of the body; it just seems so from the point of view of mind. However, all there is to mind is awareness. Mind is the limited form or activity that awareness itself freely assumes in order to know objective experience. In the *neti neti* process, the mind brings itself to the recognition that the knowing with which it knows its experience – pure awareness – is not mixed with or limited by any of the objects that it knows or is aware of. In the *neti neti* process, awareness ceases veiling itself with its own activity and paves the way for the recognition of itself as it is, eternal and infinite, just as the recognition that the screen does not share the limits or destiny of any of the characters in a movie paves the way for the recognition of the screen as it essentially is.

In its native condition awareness has no objective qualities, for any objective quality has a form, and anything that has a form is *known*; it does not know. Awareness is the non-objective *knower* of all objective knowledge and experience: it is pure knowing or being aware. Being without form, awareness is said to be empty, that is, empty of objective content. It is the witness of all objects but is not itself an object of knowledge or experience.

The discovery that awareness is the witness of all objective experience partially liberates it from the limitations of the body in and as which it seemed to be located, but it is not a full liberation. The discovery 'I am awareness or consciousness itself' is not what is referred to as enlightenment or awakening, although it is often mistaken as such in contemporary expressions of the non-dual understanding.

At this stage of understanding, although awareness is no longer exclusively identified with the body-mind, it still seems to be located and limited. As such, it still seems to be temporary and finite, the two beliefs that give rise to the fear of death and the sense of lack around which the apparently separate self or finite mind revolves. To fully liberate awareness from all superimposed beliefs and feelings, a further investigation is required as to the essential nature of awareness itself.

CHAPTER 5

THE DIRECT PATH TO ENLIGHTENMENT

Having discovered that the experience of simply being aware or awareness itself is our primary, essential and irreducible identity, the question remains as to the *nature* of awareness or consciousness itself. The next four chapters detail the process by which the nature of consciousness is explored and revealed.

Science and psychology focus their investigation exclusively on the phenomenal aspects of experience, mapping the known world with ever-increasing precision. Science is defined by the *Oxford English Dictionary* as 'the intellectual and practical activity encompassing the systematic study of the structure and behaviour of the physical and natural world through observation and experiment'. Psychology is defined as 'the scientific study of the human mind and its functions'.

All that is or could ever be known of the 'physical and natural world' are sensations and perceptions, and of the 'human mind and its functions', thoughts, images and feelings. In other words, all that is or could ever be known of the physical world and the human mind is experience, and all experience takes place in consciousness. Therefore, using the same rigor and honesty by which science is defined, the disciplines of science and psychology could be redefined as the study of the *contents* of consciousness. In neither case is consciousness *itself* studied.

Why not? Because scientists and psychologists, along with the majority of humanity, have limited their interest and observation to the study of *objective* experience, thereby overlooking and, in some extreme cases, altogether denying the *subjective* experience of consciousness or awareness itself. Consciousness is considered, if it is considered at all, to be a by-product of its contents. Anyone would be ridiculed for believing that the sky is produced by clouds, that television screens are created by movies or that oceans are generated by waves, but such is the extraordinary degree to which our culture has departed from reality that when applied to the matter of experience itself, such an idea is revered. Indeed, it has become, without our even realising it, the very foundation of our world culture.

From the perspective of science, objective experience consists of the body and world, and subjective experience – the realm of psychology – of thoughts, images and feelings. Thus, the objective realm is equated with public or *shared* experience, the subjective realm with *private, idiosyncratic* experience. The world is considered shared and consciousness private, and therefore only the world is considered a worthy subject of empirical study.

However, from the consciousness-only perspective, both our experience of the body and world *and* our experience of thoughts, images and feelings are considered objective – not because they are shared but because both have observable qualities or characteristics. Consciousness alone is considered subjective, that is, the subjective knower of all objective knowledge and experience. However, it is also legitimate in this context to say that all experience is subjective, meaning that all experience takes place within our own minds. In each of these statements the words 'objective' and 'subjective' are used with different meanings, and as long as these meanings are understood there is no contradiction between them. What is important is that whichever formulation is used, experience not be divided into two essential realms: it is either all subjective or all objective.

Let us stick to the former meaning, in which all experience is considered objective and consciousness alone subjective. Being subjective and without form or observable qualities, the experience of being aware, awareness or consciousness itself is overlooked by most people in favour of objective experience. This approach is taken to an extreme in the field of Consciousness Studies. Instead of expanding the parameters of science to include the non-objective experience of being aware, researchers have superimposed the limitations of their own minds onto consciousness itself, trying thereby to bring consciousness within the limited parameters of objective experience, and thus reducing the study of consciousness to the study of neuroscience. However, neurology can tell us nothing about the *nature* of consciousness, nor indeed can any other branch of science. Only consciousness knows consciousness.

Everyone has equal access to the experience of being aware, awareness or consciousness itself, and therefore everyone is equally qualified to investigate its nature. One does not need to be a trained scientist or psychologist, nor does one need to belong to any religious or spiritual tradition. In fact, the preconceived ideas and, more importantly, the emotional investment that may attend such affiliations often preclude a rigorously honest and impartial investigation of experience, leaving the field open to free thinkers outside the constraints of any formal institution.

*　　*　　*

How are we going to study consciousness or awareness? Before being able to answer that question we have to decide *what* is going to study awareness. Only that which knows a thing can tell us anything about that thing. Therefore, only that which knows awareness can tell us anything about awareness. What is it that knows awareness or the experience of being aware? Thoughts, images, feelings, sensations and perceptions are not *themselves* aware; they are objects we are aware *of*. But who is the 'we' that is aware of them? It is awareness itself, the knowing element in all experience. Only awareness is aware, and therefore only awareness knows anything about awareness.

The study of awareness is a unique science in that it is entirely between awareness and itself. In the East it is known as meditation; in the West it is known as prayer. Although the practice of meditation has been reduced by popular culture to a means of relieving stress and anxiety, in its original form it is the means by which awareness has access to its knowledge of itself. Likewise, although through the ages the practice of prayer has become the means whereby a personal self supplicates an external god, in its original meaning it was understood to be the means by which the finite mind surrenders its limitations and is revealed to itself in its original, irreducible form.

Thus, in the original meaning of the words, meditation and prayer are the means by which the mind has access to its own reality of eternal, infinite awareness. However, in our age the disciplines of meditation and prayer have for the most part been mixed with the local, temporal traditions in which they are practised and, as a result, their essence has been obscured or lost in favour of exoteric spiritual and religious practices.

In other cases, the study of the nature of awareness itself was considered too esoteric or difficult for those whose minds were exclusively involved with and fascinated by objective experience. Therefore, as a concession to the object-facing mind, a long and arduous series of preparatory practices was considered necessary to purify the mind of its outward-going tendencies before it was considered fit for the ultimate endeavour of exploring its own reality.

As a result, the world's religions have evolved into a mixture of the essential, irreducible reality of all experience – awareness itself – and local, temporal traditions, the former becoming progressively obscured by the latter, to such an extent that the world's major religions have now, to a greater or lesser degree, lost touch with the understanding from which they originated.

However, this understanding has always been preserved in remote ashrams and monasteries, and in the hearts and minds of individuals who deeply desire to know the nature of themselves or the ultimate reality of the world. Three such individuals were Ramana Maharshi, Atmananda Krishna Menon and Sri Nisargadatta Maharaj, who, in the middle of the twentieth century, between them resurrected the means by which the nature of awareness might be discovered by ordinary people without any prior knowledge or preparation and with no particular religious or spiritual affiliation.

These three sages did not invent the means by which the essential nature of reality may be recognised; they simply extricated and resurrected it from the shroud of accumulated belief by which it had been obscured for centuries, divesting it of the exotic cultural packaging which served only to mystify it, and reformulated it for a new generation.

This approach is sometimes referred to as the Direct Path, in reference to the fact that we go directly to the recognition of our essential nature of pure awareness. That is, awareness goes directly to the knowing of itself without having to direct itself towards any objective practice or experience. This is in contrast with many traditional or progressive approaches, which demand a succession of practices as a prerequisite for this understanding, all of which require the directing or focusing of awareness towards an object, such as a mantra, a flame, the breath, the pause between breaths, the guru, and so on, in order to purify it of its accumulated conditioning.

Whilst there may have been justification in previous ages for more progressive approaches to the reality of experience, it is my opinion that our age is ripe for the Direct Path. The resistance to the Direct Path felt by some people who have been on a progressive path for many years or even decades is not due to its difficulty or unsuitability. It is rather that such people have invested themselves in and become attached to and limited by the very forms and practices which, ironically, they initially undertook precisely for the purpose of freeing themselves from all such limits.

In the Direct Path it is recognised that the experience of being aware or awareness itself is the knowing element in all experience, irrespective of the content of experience, and thus no *particular* experience is a carrier, indication or hallmark of awareness itself. Just as the screen is equally evident in all movies, irrespective of their content, so awareness shines equally brightly in all experience, from our deepest depressions to our most joyful feelings.

The Direct Path requires no spiritual or religious affiliation or background. No preparatory practice is required for the simple recognition that being aware or awareness itself is the essential, irreducible, indivisible element in all experience, to which we refer as 'I' or 'myself'. None has privileged access to it. It is equally available to all people, in all places, under all circumstances and in all situations. All that is required is a deep interest in the nature of reality, or an intuition that the peace, happiness or love for which all people long cannot be satisfied by objective experience.

*   *   *

How is consciousness going to discover its own nature? In order to know objective experience, consciousness must assume the form of mind, but it cannot know *itself* in the form of mind. The world is known by the mind; the mind is known by consciousness; and consciousness is known by itself.

As consciousness is the only 'one' that can know itself, we have to ask consciousness itself, as it were, what *its* own knowledge or experience of itself is. Thought can formulate the question, 'What is the nature of the knowing with which experience is known?' but only that which *knows* experience can legitimately answer it. All experience is known by consciousness, and therefore only consciousness can know consciousness. Only awareness can be aware of being aware.

I recently attended a talk on the nature of consciousness given by a well-known and highly respected professor of philosophy at Oxford University. In the midst of the abstract and convoluted lines of reasoning with which he attempted to describe the nature of consciousness, he made the remark, 'Some philosophers say it is possible for consciousness to be aware of itself; such ideas should be put in the trash.'

At the end of the meeting, I said to him, 'Everybody in this room is aware that they are conscious.' He pondered for a moment and then agreed. I then observed that whatever it is that is aware that we are conscious must itself have two qualities: one, it must be present, and two, it must be aware. 'What would that be?' I asked rhetorically. 'Consciousness! Consciousness is precisely that which is present and aware.' I suggested that the experience of simply *being aware* that each one of us was having at that very moment *is* itself the experience of consciousness knowing itself.

'Oh, no, no', he said, 'you're going far too quickly', and took off again along lines of elaborate and abstruse reasoning, which bore no relation to experience and which concealed his ignorance in obscurity and complexity. When I remonstrated with him again, he simply turned away and took the next question. It was an example of profound ignorance masquerading as understanding, all the more shocking for coming from an eminent professor at one of the world's top educational establishments.

The professor had overlooked the simplest, most intimate and fundamental experience: the knowing of our own being, the consciousness of existence. If consciousness did not know its own being, we would not know the experience of being aware and would, as a result, have to answer 'no' or 'I don't know' to the question 'Are you aware?' The experience of being aware is everyone's primary experience, although it is usually overlooked in favour of objective experience. When I say it is 'everyone's' experience, I mean it is awareness's primary experience that it is aware.

Consciousness's knowing of its own being shines equally in all people, although in most cases we seem to lose our essential identity to objective experience. Instead of being aware of the feeling of sadness, we feel I *am* sad. Instead of being aware of the feeling of loneliness, we feel I *am* lonely. Instead of being aware of the thoughts, feelings, sensations and perceptions that accompany the ageing process, we feel I *am* getting old. In this way we allow our essential being, and with it its innate qualities of freedom and peace, to be appropriated and thus conditioned by experience.

The knowledge of our own being – the experience that shines in the mind as the knowledge 'I am' – is the same in all people. None has privileged access to it, and for this reason it is the only knowledge about which it is not possible to disagree. It is absolutely true, under all states, circumstances and conditions. If we find ourselves disagreeing about the nature of our essential being then we are mistaking ourself for an object. It is only possible to disagree about something that has objective qualities.

If Ramana Maharshi and Hitler were both to refer to their essential feeling of being, they would both refer to the *identical* experience. It is true that in the former case, the feeling of being would shine as it is, unlimited or uncoloured by any of the characteristics of the body or mind, whilst in the latter it would seem to be mixed with the qualities of the body and mind and, as a result, appear to share their limitations and destiny. In the ultimate analysis, only the clarity with which their own being were known to each of them would account for the difference between their ideas and feelings, and the subsequent expression of them in their activities and relationships.

Just as one person watching a movie will be so involved in the drama as to see only a multiplicity and diversity of separate and conflicting objects and selves, whilst another, watching the same movie, will see only a single, indivisible and inherently 'peaceful' screen, so the self-aware 'screen' of consciousness, upon which all experience plays, is equally available to all people in all circumstances, though it will appear differently according to the mind through which it is filtered and the form in which it seems to be modified or coloured.

To say that consciousness is equally available to all people suggests that *people* are aware of or have access to consciousness. That is just a manner of speaking. Only consciousness knows and, therefore, has access to itself. Sometimes consciousness colours itself in the form of the mind and seems, as a result, to become finite and temporary. This is known as the states of waking and dreaming. At other times consciousness ceases to colour itself and remains in its uncoloured condition, knowing only its unlimited, inherently peaceful being. This is the experience of deep sleep.

In both cases consciousness is only knowing itself. From the perspective of the waking-state mind, consciousness seems to be limited or finite in the waking and dream states, and to be missing or absent in deep sleep. However, the states of waking and dreaming are consciousness knowing itself in the form of the temporary, finite mind, and deep sleep is consciousness knowing its own eternal, infinite, unconditioned being. In other words, the 'somethingness' of the waking and dreaming states and the 'nothingness' of deep sleep are superimposed on consciousness by thought, making it seem that consciousness itself assumes the forms of three separate states.

The mind that believes that consciousness is limited is like water believing that its essential form is a wave; the mind that believes that consciousness is absent is like a wave believing that water is missing. Consciousness's knowing of its own being is like water simply knowing itself as water, irrespective of its form as a wave. In fact, water knows nothing of waves; it knows only water. Likewise, consciousness knows only itself.

*　　*　　*

Everything the mind knows, with one exception, is founded upon the belief that consciousness shares the destiny and the limits of the body. Thus, all human knowledge, with one exception, is untrue, or at least only true relative to the perspective of a temporary, finite consciousness.

However, the *knowing* with which the finite mind knows its limited knowledge is infinite consciousness, refracted through the lens of its own activity. So although everything that the mind knows is at best only relatively true, at the same time all knowledge is founded upon and made of the light of absolute truth. That is why William Blake said, 'Everything possible to be believed is an image of truth.'*

This light of pure knowing, which shines in the mind as the knowledge 'I am', and which is the ultimate foundation upon which all the mind's relative and limited knowledge is based, is the 'one exception' by which the mind has access to reality.

The understanding that everything known by or through the mind is untrue or at least only relatively true will, in many cases, precipitate a crisis. In my case this crisis appeared spontaneously when I was seventeen. Whilst still at school studying science with the intention of becoming a biochemist, I was struck by a realisation that was to change the course of my life.

One day, shortly after the end of a physics lesson, I was sitting on my bed, my mind simply open and available. Into this availability a thought

---

* From *The Marriage of Heaven and Hell* (c. 1790).

appeared as if from nowhere, unsolicited by any line of reasoning in which I was currently engaged or had ever pursued: 'There can be no end to objective knowledge.'

I realised that the mind could only ever know its own content and that all such content, being objective, was limited. It became clear in an instant that all thought would evolve, change and eventually disappear, and could not therefore be relied upon as being absolutely true. A presentiment began to form itself in my mind, although it was many years before it was clearly formulated: 'What can we know for certain that is absolutely true? Can anything that is known by the mind be relied upon?'

As the thought developed, I noticed that it was accompanied by a growing feeling of fear and dread in the body, a feeling that everything in which I had invested my identity and beliefs was subject to change and could not be reliable or certain. I felt that the solidity of the universe was dissolving around me. It was in response to the discomfort of this feeling that I turned away from the intuition and sought distraction and relief in objects and activities.

However, such a thought is like an minor earthquake, the effects of which are not always immediately discernible on the surface. Although I didn't realise it at the time, this recognition was to be the first crack in what had been until then the unassailable edifice of Western materialist education.

In my case this realisation put an end to my incipient career as a scientist and initiated an investigation into the nature of beauty, which I would explore as an artist in my studio for the next thirty years. However, although I could not see it at the time, the scientist's search for knowledge is exactly the same search as an artist's quest for beauty. In fact, the scientist's search is not for knowledge itself but for understanding. Knowledge is thought, but understanding takes place when thought comes to an end, just as beauty is revealed at the end of a perception.

Understanding is thus the *end of knowledge*. It is, of course, no coincidence that the Sanskrit word *Vedānta* comes from *veda*, meaning 'knowledge', and *anta*, meaning 'end', and therefore means the end of knowledge. The beauty of language is that so much understanding can be expressed with so few words. Would that I possessed that skill! In those ten minutes at school I had spontaneously intuited the end of knowledge, although it would be another twenty years before I realised that the end of knowledge and its source or beginning are the same.

If everything that is known by or through the mind is a reflection of the assumption that consciousness is limited – a belief that, when explored, is found to be untrue – can the mind know anything that is absolutely true? The simple answer is 'No', for the mind itself is a limited form of consciousness, whose apparent existence as a separate entity, and whose entire body of knowledge, are only true from its own limited point of view.

So we could say that everything the mind knows is a reflection of its assumption of itself as a separate entity – the temporary, finite consciousness that knows itself as 'I, the body' – that lies at its origin. In the same way that a screen assumes the form of an image with which it seems to veil itself, so eternal, infinite consciousness assumes the forms of thought and perception, from whose perspective consciousness itself seems to be temporary and finite.

Thus, all the mind's knowledge and experience appear in a form that is consistent with its own limitations, including its own belief in itself as an independently existing self or entity, just as the character in a dream, who in its own view of itself is an independently exiting *entity*, is in fact simply the *activity* of the dreamer's mind. At the same time, the very knowing with which the separate self knows its apparently finite experience, irrespective of however good, bad, right, wrong, pleasant or unpleasant that experience may be, is itself the eternal, infinite light of pure knowing or consciousness, whose knowing of its own being is the only absolutely true knowledge there is.

It is the infinite light of pure knowing that precipitates within itself a multiplicity and diversity of finite knowledge and, in so doing, seems to veil the knowing of its own being from itself. As a result, the true and only reality of pure consciousness now seems to experience itself as a temporary, finite self that shares the destiny and limits of the body. With this belief, the peace, happiness and love that are reflections of the knowing of our own being are apparently lost, though in fact only ignored, overlooked or forgotten.

This veiling of peace, happiness or love is like a wound which resides in the heart of the apparently separate self, and which continuously tries to alleviate itself through the acquisition of objects, substances, activities, states of mind and relationships. However, at some point, either through intuition or at the suggestion of a friend, it becomes clear that this longing will never be fulfilled by objective experience. As a result, a new possibility opens up, in which it is recognised that in order to fulfil its longing for lasting peace and happiness the mind must be divested of its limitations.

All spiritual and religious paths, in one way or another, are aimed at divesting the mind of or expanding it beyond its limitations. However, when the mind is relieved of its limitations it ceases to be mind, as such, and stands revealed as the eternal, infinite consciousness that is its essential, irreducible reality. That knowing of our own being as it truly is – consciousness's knowing of itself in us – *is* the experience of peace, happiness or love.

CHAPTER 6

What does consciousness have to do to know itself? In order for consciousness or awareness to discover its own nature, it has to 'look at' itself. However, awareness cannot separate itself from itself and look at or know itself in subject–object relationship, just as the sun cannot turn round and shine on itself.

To know objective experience awareness assumes the form of mind, but to know *itself* awareness need not assume the form of mind; it need only remain in itself. This self-resting in and as awareness is the essence of meditation or prayer. It is the non-activity in which awareness knows its own being.

A mind that is accustomed to directing its attention exclusively towards objective experience will often object to this suggestion, saying that it does not know where to find awareness or in what direction to look. For such a mind the presence of awareness may first be accessed as the experience of simply being aware, the feeling of being or the knowledge 'I am'. Later on, the mind will recognise that all there is to experience is the knowing of it, and therefore the only substance that awareness ever truly knows or comes in contact with is itself.

However, a mind that has been indoctrinated with the materialist paradigm from an early age will believe that it experiences a multiplicity and diversity of objects, each with its own separate and independent existence, which it seems to know from the perspective of a separate subject or self, just as the dreamed world seems to be known from the perspective of a dreamed subject *within that world*. Thus, as a compassionate concession to such a mind we may say, to begin with, that the presence of awareness is known as the experience of simply being aware, the feeling of being or the knowledge 'I am'. Turning the mind towards any of these will take the mind on a unique journey, in a directionless direction, on a pathless path, in which it will gradually dissolve into the light of awareness from which it arises, like an image slowly fading, leaving the screen, its reality, visible.

In religious terms this resting of the mind in its source is referred to as the practice of the presence of God. The penultimate prayer is the directing of the mind exclusively towards God, but this prayer still leaves the mind intact as an entity separate from God. The ultimate prayer is the surrender of the very mind which considers itself to be an entity in its own right. To the mind that believes itself to be a limited entity in its own right, God is something other than itself. But if God is something other, then God must be finite, and such a God cannot be God. As the Sufi mystic Balyani said, 'Otherness for Him is Him without otherness.'

In other words, God can only be known by Himself, and in order to know His own existence God must be self-aware. Therefore, God's knowledge of Himself must be awareness's knowledge of itself. Awareness's knowledge of itself shines in each of our minds as the knowledge 'I am', the feeling of being or the simple experience of being aware. That is, the knowledge that each of us has of our very own being is God's knowledge of Himself. Later we will also see that the ultimate reality – awareness's knowledge of itself or God's knowledge of Himself – shines in the experience of peace, happiness, love and beauty and, in fact, expresses itself in and as the totality of all experience.

Thus the knowledge 'I am' is God's signature in the mind. It is the portal through which awareness localises itself as the mind and the same portal through which the mind passes in the opposite direction as it investigates its essential nature. The knowledge 'I am', or the knowledge of our own existence – awareness's knowing of its own being – is our primary knowledge, upon which all other knowledge and experience depend. Until the nature of ourself is known, it is not possible to have correct knowledge about any other thing. Thus, there is no higher knowledge than to know the nature of oneself, the nature of 'I'.

We may not know exactly *what* I am, but we know *that* I am. Before we know anything about ourself, such as our age, name, gender, nationality, height, weight and qualifications, each of us knows that 'I am'. That is, before awareness knows any objective knowledge or experience, it knows its own being. In the experiences 'I am young', 'I am old', 'I am sick', 'I am well', 'I am a man', 'I am a woman', 'I am sad', 'I am happy', and so on, the simple knowledge of our being – the knowledge 'I am' – remains consistently present, although it may seem to be temporarily coloured by experience.

Our essential, self-aware being doesn't appear or disappear. However, the mind confuses all temporary qualities, conditions, ideas, images and feelings with the basic feeling of being and thus imagines that our essential being shares their limitations and destiny, and is, as such, temporary and finite. The knowledge 'I am' never changes, although it sometimes seems to be obscured, veiled or coloured by experience. The states of waking, dreaming and sleeping and the qualities, conditions, ideas, images and feelings that accompany them are temporarily added to our basic being but never essentially define or modify it.

Unlike all other knowledge and experience that is known by 'I', the knowledge 'I am' is known by itself. It is *I* that knows that *I am*, or, as God said to Moses in the Old Testament, 'I am that I am.' The knowledge 'I am' is a trace in the mind of the vast ocean of awareness that lies beyond and prior to the mind, and indeed *in which* the mind appears. In the same way, a patch of blue sky that at first seems to appear within the clouds is itself a hint of the vast sky that lies beyond and prior to them, and indeed *in which* they are located.

The knowledge 'I am' is the only knowledge that remains the same under all circumstances, at all times and for all people, and is thus a hint in the finite mind as to its essential, irreducible reality. No other knowledge satisfies these requirements, and therefore no knowledge other than awareness's knowing of its own infinite being can be said to be absolute. This knowledge is known in religious terms as the Absolute or God's infinite being. The knowledge 'I am' is the first form of God in the finite mind. 'I am' is thus said to be God's holy name. God Himself has no name, but in the mind He shines as the name 'I' or 'I am'.

*   *   *

At this stage, the mind may legitimately ask how it can turn the light of its knowing away from the objects that it seems to know and towards its own essence. This turning around of the mind is not an activity that the mind can undertake, but rather the cessation of an activity of which it was previously unaware. However, as a concession to the mind's belief that this turning around is something it can do, the teaching may now elaborate a process in which the attention that the mind normally gives to objects that are conceived as *other* than itself is instead turned around and focused upon *itself*, the subject or knower.

All objects of knowledge and experience are known from the perspective of the finite mind, the apparently separate subject of experience, and everything the mind knows or experiences is a reflection of and appears in accordance with its own limitations. Therefore, in order to know what anything truly is, the mind must first turn the light of its knowing upon itself. That is, before attending to objects so as to know what they truly are, it must first attend to itself to know what *it* truly is. The mind must shine the light of its knowing away from the objects it seems to know, towards itself, towards the very knowing with which it knows its knowledge. Attention must attend to attention itself.

The nature of the knowing or awareness with which the mind knows its knowledge is simply to be and to know. It is for this reason that it is called *pure* knowing or *pure* awareness, in the sense that its knowing is not mixed with anything other than itself; it is like the colourless light in all colour. Therefore, in knowing itself the mind does not acquire any new knowledge, in the way that it might discover a new species, mathematical equation or recipe. The realisation of the nature of the knowing or awareness with which the mind knows its knowledge is not a new form of objective knowledge. Rather, it is the remembering, recognising or knowing again of the pure knowing that is seemingly veiled, forgotten or overlooked as a result of the mind's focusing on objective experience.

Thus, to know itself as it is, the mind need only relax the focus of its attention from the objects that it seems to know and allow its knowing to fall or flow back into itself. In fact, it is not so much that the mind *focuses* on things that are 'other than itself', but rather that it becomes *mixed with* or *lost in* its knowledge of things, in the same way that a screen seems to get mixed with or lost in the movie.

The mind's knowledge and experience are never separate from itself. The mind never knows anything at a distance from itself, just as the movie never takes place at a distance from the screen. The self-aware screen does not need to separate itself from the objects or characters in the movie to realise it doesn't share their limitations or destiny. Indeed, it cannot separate itself from them, because they are only a modulation of its own being. All that is necessary is for the screen to simply 'see' that its essential nature is not inherently limited by the forms that it assumes.

Likewise, the mind does not need to separate itself from or reject any experience. It need only understand that the knowing with which it knows its knowledge and experience is already and always inherently free of anything that it knows or experiences.

When the knowing or pure awareness with which all knowledge and experience are known gives its attention to itself rather than to objective experience, its essential, irreducible nature, which previously seemed to be obscured or veiled by objective experience, seems suddenly to shine as it is. In fact, it was always shining as it is, but it was previously mixed with and thus coloured by objective experience and, as a result, seemed to be missing or obscured.

*   *   *

What is described as the turning around of the mind upon itself is not so much a 'turning around' of the mind as a sinking, falling back or relaxing into itself. The phrase 'turning around' is used only in contrast and as a concession to the previous, object-facing direction of the mind. To a mind that is facing outwards to know objects, the non-dual teaching suggests turning itself around towards its source or essence in order to know itself.

However, anyone who has ever tried to turn their attention around in this way knows that it cannot be done. The mind cannot 'turn around' towards the objectless source of its own knowing, for it can only know or be directed towards an object. Anything the mind turns towards, even if it turns in the opposite direction of the objects or thoughts that it normally knows, would necessarily be in the direction of another object. Thus, in the same way that one cannot stand up and take a step towards oneself, so the mind cannot turn around and direct itself towards its own source.

The suggestion to turn the mind around upon its source is a concession to the separate, mind-made self – the temporary, limited and ultimately illusory form of awareness that knows itself in each one of us as the belief and feeling 'I am the body' – and to its belief that it is a real entity that is in control of its own destiny.

This turning of the mind around upon its source is sometimes known as self-enquiry, which is a translation of the Sanskrit term *atma vichara*. However, self-enquiry is a unique kind of investigation which does not involve an exploration of any kind of objective knowledge but rather investigates the subjective knower of all objective knowledge and experience. As such, the term 'self-enquiry' is, again, at best a concession to the mind that initially imagines it can explore its own nature in the same way that it explores objects.

To such a mind the teaching suggests enquiring into its own nature, an investigation that may be initiated by a thought such as, 'What is it that knows or is aware of my experience?', 'What is the nature of the knowing with which all knowledge and experience are known?' or 'Who am I?' However, this investigation does not require the mind to direct itself towards any kind of objective knowledge or experience. It is rather a falling, sinking or relaxing back of the attention, or the focus of the mind's knowing, into its source.

The meaning of the word 'attention' gives us a hint as to the nature of this investigation into our true nature. The word 'attention' comes from the Latin *attendere*, a compound of *ad-*, meaning 'to' or 'towards', and *tendere*, meaning 'to stretch'. Thus, attention is a stretching of the knowing that is the essence of the mind towards a thought or object.

When we know a thought, the knowing with which that thought is known proceeds from its source towards that object. When we know a sound, the same light of knowing proceeds from the same source towards the sound. When we attend to any object – whether it seems to be something within ourself, such as a thought or feeling, or something outside ourself, such as a sight or sound – the light of knowing proceeds from our self, the subject, towards that object.

This experience of being a subject that knows or attends to an object is expressed in our language in conventional dualistic terms such as, 'I know a thought'. That is, 'I', the subject of experience, shines its light of knowing towards the object that it knows. Our attention or light of knowing is stretched from its source, the subject 'I', towards its object. But what happens to attention when it no longer has anything to stretch itself towards? What happens when a piece of elastic that is stretched between two points is released from one of those points? Having nothing else to attach itself to, it springs back to its original, unstretched position.

In most people attention moves perpetually from one object or experience to another, with only brief periods of respite in the interval between two thoughts or perceptions, in moments of peace or happiness, and during deep sleep, when the mind is at rest in its source. But when, through interest in its own essential nature, the mind ceases to direct itself towards objective experience, it begins to sink or relax back into the source from which it has arisen. This source is pure knowing or awareness, before it knows or becomes mixed with any objective knowledge.

This falling back of the attention into its source is the means by which the mind comes to know its original nature. In fact, this is not an *activity* of the mind, although it may seem as such to begin with, but rather the *cessation* of a previous activity – the activity of overlooking the knowing of its own objectless, self-aware being in favour of objective knowledge and experience.

The falling or sinking back of the mind into its source is the essence of meditation and prayer and can be found, in one form or another, in all the great religious and spiritual traditions. Ramana Maharshi referred to it as the sinking of the mind into the heart of awareness. Nisargadatta Maharaj referred to it as focusing on the experience 'I am'. The Russian philosopher P. D. Ouspensky referred to it as self-remembering. The poet Tennyson suggested seeking the ultimate nature of the mind as one would follow a 'sinking star, beyond the utmost reach of human thought'.*

Referring to the same non-practice, St. Matthew tells us, 'When thou prayest, enter into thy closet, and when thou hast shut thy door, pray to thy Father which is in secret; and thy Father which seeth in secret shall reward thee openly.' The Kashmir Shaivite mystic Lalla referred to this turning around of the mind when she said, 'I have travelled a long way seeking God, but when I finally gave up and turned back, there He was, within me.'

*   *   *

To better understand the return of the mind to its source of pure awareness, let us take a new metaphor. Imagine an actor called John Smith who leads a fulfilled and happy life. One day John Smith is assigned the role of King Lear, which involves learning a set of lines and putting on a costume. On the first night of the performance, John Smith leaves home and goes to the theatre. He dresses up in King Lear's clothes and assumes

---

*Alfred, Lord Tennyson, 'Ulysses' (1833).

his thoughts and feelings. John Smith is an excellent actor, so he completely forgets his own thoughts and feelings and, as it were, becomes King Lear.

The performance begins and John Smith now thinks, feels, acts, perceives and relates as King Lear. As the play develops, John Smith becomes more and more unhappy, thinking that everyone and everything is against him. In fact, so completely does John Smith forget his own nature that when the performance ends, he forgets to take off King Lear's words and clothes when he returns to his dressing room.

Several friends come into his dressing room to congratulate him on his performance and, finding him utterly miserable, try to persuade him that he has forgotten who he truly is, but to no avail. After a few minutes a friendly stranger comes in and sits beside him, and they start conversing. The friendly stranger asks him to tell him about himself, and King Lear starts to describe himself: 'I am King of England. I am eighty years old. I am the father of three daughters.'

King Lear goes on to describe the problem of dividing his kingdom between his three daughters, and about the conflicts in which he is embroiled. The friendly stranger asks him to keep going. King Lear begins to describe his thoughts and feelings: 'I am intelligent, kind, confused, lonely, restless, anxious...' The stranger asks him again to keep going. King Lear begins to falter as he tries to find words to describe the subtler qualities of himself. Pauses begin to appear in his description as King Lear reflects back through deeper and deeper layers of himself.

Without realising it, King Lear has ceased giving attention to the thoughts, feelings and circumstances with which he is usually occupied and is instead giving attention to that part of himself that lies below them, a nameless, formless well of barely discernible feelings. 'There is a dissatisfaction in me, but I am not sure exactly what I am dissatisfied with. There is a longing in me, but I am not sure what I am longing for.'

The friendly stranger remains silent, which King Lear takes as an invitation to continue. He keeps going deeper into himself until he can go no further. There is a long silence. 'I am... I am...' Nothing follows. He waits. From time to time a memory of his daughters disturbs him, but so interested has he become in discovering who he essentially is that he gives the memory no attention, and in due course it leaves.

'I am what?' the stranger asks, to help King Lear further focus the essential feeling of being in his mind.

'I am... I am... There is...' A long silence follows.

King Lear is experiencing the most essential, irreducible element of himself – unqualified being – before it is mixed with any thoughts or feelings. He cannot describe it because none of the words he is accustomed to using in relation to thoughts, feelings, sensations and perceptions are applicable. They all seem too crude for the purpose.

He remains in silence for some time without the need or ability to describe his experience. When a thought arises, it is only to comment on the unusual and yet, at the same time, strangely familiar peace that he now feels. It is a peace that has no obvious cause in his external circumstances. It is prior to and independent of anything that he previously believed himself to be. He remains silent, and he has no idea how long he remains at rest in this way, for thought is no longer active and therefore he cannot take account of time.

At some point it occurs to King Lear that the peace he now feels has not been added to him as a result of anything that is or is not taking place in his life. The friendly stranger did not give it to him, nor can he take it away from him. It seems to come from within himself. He realises that the peace he now feels is, in fact, always present in the depths of his being. He recognises that his own inherently peaceful being is always available, lying just behind or underneath the turbulent flow of his thoughts and feelings, and is independent of the drama of his life and the conflicts in his relationships.

King Lear hears a voice inside himself say, 'Turn towards me and I will take you into myself.' He realises that it is his own deepest intelligence that is speaking to himself. He recalls the words of Isaiah in the Old Testament, 'Thou wilt keep him in perfect peace whose mind is stayed on Thee.'

He sits in silence again, allowing the full implications of this recognition to unfold in his mind. 'If this peace is inherent in what I essentially am, then I must take it with me wherever I go. This peace cannot be disturbed by my changing thoughts and feelings or the troubling circumstances in which I find myself.' An inner smile lights up inside him and an immense joy – an ancient joy that he knew as a child and yet is now experiencing as if for the first time – floods his being. It is the joy of recognising his essential, unassailable being, its freedom, its availability, its imperturbability, its unconditioned nature.

At that moment he remembers he is John Smith. This remembering of himself is not a memory of something that he once knew in the past and subsequently forgot, but a memory of something that lies ever-present within him and that was simply obscured or ignored due to his fascination with and absorption in the life of King Lear. He looks around and realises that the friendly stranger has disappeared.

* * *

John Smith ponders King Lear and the friendly stranger who came to him: 'In order to become King Lear I had to forget myself, and this forgetting caused me pain and anguish. This pain initiated a great search for peace and happiness, and such was the intensity of the search that the latent intelligence of my own mind appeared in the form of the friendly stranger who asked me about the essential nature of myself.'

John Smith realises that King Lear was simply a self-assumed limitation of his own being. This self-forgetting enabled John Smith to assume the character of King Lear. King Lear and John Smith think, feel, act, perceive and relate in different ways, and yet they are essentially the same person. The 'I' of each of them is the same 'I'. The essential self in King Lear is the same as the self of John Smith. King Lear does not have his own self. The 'I' is common to them both. The 'I' of King Lear is the 'I' of John Smith, with an imaginary, self-assumed limit attached to it. All King Lear had to do was to question the nature of this 'I'.

When John Smith recognises or remembers he is John Smith, he wakes up as if from a dream. But King Lear did not suddenly *become* John Smith when he woke up. King Lear was never King Lear. He was always only John Smith with a self-assumed limit. John Smith simply ceased imagining that he was King Lear. John Smith realises that he is always only John Smith but that his essential, irreducible being had become so mixed up with the thoughts and feelings of King Lear as to make it seem as if he had forgotten who he was. And with the forgetting or overlooking of his own being, its innate qualities of peace, fulfilment and love were eclipsed. However, John Smith did not cease being John Smith and become King

Lear, and King Lear did not cease being King Lear and become John Smith. King Lear was simply an imaginary, self-assumed limitation on the true and only self of John Smith.

Only from King Lear's illusory point of view was he King Lear. Believing himself to be King Lear, he found himself miserable, and having failed to secure the peace and happiness for which he longed in his circumstances and relationships, he turned within at the suggestion of a friendly stranger. Without realising it King Lear was spontaneously engaging in what would later be formalised as a practice of self-remembering or self-enquiry: a sinking or relaxing of his attention through deeper and deeper layers of himself until he could go no further, that is, until he arrived at his own unqualified and inherently peaceful being.

There is another long pause as John Smith ponders his experience: 'If King Lear's mind was a limitation of my own mind, could it be that my *own* mind is a self-assumed limitation of a greater, unlimited mind, which is, at the same time, who I essentially am?' John Smith begins to contemplate the nature of his own mind. Without realising it, he too is now spontaneously practicing self-enquiry.

John Smith sits quietly in his chair after everyone else has gone home. Half an hour passes by without his noticing, during which time his mind sinks more and more deeply into itself, and then, suddenly, quietly, he realises that even John Smith is a self-assumed limitation on his essential being. As if from nowhere, he feels a flood of peace and joy. He opens his eyes and notices that the room too is saturated with peace. He smiles to himself, remembers his family with great love, and walks home.

CHAPTER 7

THE EXPERIENCE OF BEING AWARE

Imagine asking an English, Chinese, Ghanaian or Aboriginal person, a homeless, wealthy or sick person, none of whom had ever heard of non-duality, 'Are you aware?' As long as they all understood the question, they would all answer 'Yes'. If any of the seven billion human beings were to hear the question 'Are you aware?', they would all refer to the *same* experience – the experience of simply being aware – and would, as a result, all answer, 'Yes'.

Awareness, or the experience of simply being aware, doesn't have an English, Chinese, Ghanaian, Aboriginal, homeless, wealthy or sick flavour. It doesn't matter what our nationality, age, health, gender, wealth, education, intelligence, and so on, may be; the experience of simply being aware is the same in every case. Indeed, if a dog, chicken, fish or ant were able to understand the question 'Are you aware?' it would refer to exactly the same experience that each of us refers to.

Not only do all beings share the same experience of being aware, but each person refers to the same experience at various times in their life. When we are five years old, ten years old, twenty, forty, sixty years old, the actual experience of being aware is always the same. The experience of being aware, or awareness itself, is not conditioned or coloured by age. Moreover, for each person, the experience of being aware remains the same irrespective of what thoughts, feelings, sensations or perceptions are present.

When anyone, regardless of the state of their mind or the condition of their body, hears the question 'Are you aware?', they pause. In that pause everyone refers directly to the identical experience of being aware.

Awareness's awareness is redirected away from the object on which it was previously focused and reoriented towards itself, that is, towards the experience of simply being aware. And in doing so, everyone refers to exactly the same experience.

It is important here to make the distinction between 'similar' and 'same'. In referring to the experience of simply being aware, it is tempting at first to believe that we all refer to a *similar* experience. Such a view would suggest that there are multiple, similar awarenesses, one for each person or animal. However, if there were more than one awareness, each awareness would have to have some objective quality that distinguished it from all the others. But no such objective quality is found in our actual experience.

Thought *believes* that awareness has limiting qualities, but those qualities are never actually found in experience. That is, in awareness's own experience of itself, there are never any limitations or boundaries, just as, if space were able to look at itself, it would find no limit or boundary within itself.

When all seven billion of us refer to the experience of being aware, we refer to the *same* experience. This is difficult to imagine if we believe that awareness is located inside the brain. However, we have already seen that the experience of being aware is not qualified by any of the limitations that belong to the mind or body. Thus, in its own experience of itself, it is unlimited or infinite, and being infinite it cannot be found in time or space. In fact, in simply knowing the experience of being aware, without realising it we step out of the mind and body, and therefore we step out of time and space.

When anyone directs their attention to the experience of simply being aware and, as a result, actually *experiences* the experience of being aware, they 'go to' or 'touch' that element of experience that everyone shares or has in common. At that timeless moment – timeless because it is not found in the mind – they stand as one at the heart of humanity. The experience of love is precisely that experience, the experience of our shared being.

* * *

Everyone is aware of their experience. We may not know exactly who or what it is that is aware, nor indeed what experience actually consists of. Nevertheless, we can say for certain that we are aware of a flow of thoughts, images, ideas, feelings, sensations, sights, sounds, and so on.

If someone were to ask us, 'Are you aware of your thoughts?' we would take a moment to check our experience, directing our attention towards the current thought, and would respond, 'Yes, I am aware of my thoughts.' If someone were to ask, 'Are you aware of the tingling sensation at the soles of your feet?' we would direct our attention towards the soles of our feet, experience the sensation and reply, 'Yes, I am aware of that sensation.' And if someone were to ask, 'Are you aware of whatever sounds are currently present?' again we would direct our attention towards the current sound and respond, 'Yes, I am aware of that perception.'

In each case, in between the question as to whether we are aware of a thought, sensation or perception and our response, something takes place that enables us to respond affirmatively. What is that something? It is the *actual experience* of the thought, sensation or perception.

Now if, instead of asking whether or not we are aware of a thought, sensation or perception, someone were to ask simply, 'Are you aware?' we would pause for a moment and then respond, 'Yes'. The question 'Are you aware?' is a thought. The answer 'Yes' is another thought. In between these two thoughts something takes place which is *not* itself a thought. It is as a result of this 'something' that we are able to answer 'Yes' to the question 'Are you aware?'

In order to answer 'Yes' to the question 'Are you aware of thoughts, sensations and perceptions?' we have to *experience* the thought, sensation or perception to which the question refers. It is as a result of that experience that we are able to affirm that we are aware of each of them. To what experience do we refer in order to answer 'Yes' to the question 'Are you aware?'

The experience to which we refer to takes place in the gap between two thoughts: 'Are you aware?' and 'Yes'. What is it that takes place in the pause between these two thoughts? The experience of being aware. In that pause we become aware of being aware. Whilst everyone is aware by nature, not everyone is *aware* that they are aware. The experience of 'being aware of being aware' is triggered by the question 'Are you aware?' If we did not *experience* 'being aware' we would not be able to legitimately answer 'Yes' to the question.

Ask yourself the question, 'Am I aware?' but don't answer the question with the thought 'Yes'. Just ask the question 'Am I aware?' and allow your attention to be drawn to the experience that will subsequently inform the answer 'Yes'. The experience that takes place in between the two thoughts, 'Am I aware?' and 'Yes', is not an activity of mind. It takes place between two activities of the mind.

Stay with the pause between these two thoughts. When we remain in this pause before the answer formulates itself, what takes place 'there' is the most valuable and, at the same time, the most underrated or overlooked experience that one can have.

*   *   *

The experience of being aware is unique amongst all experiences. All other experiences – all thoughts, feelings, sensations and perceptions – have some form or objective quality to them. But the experience of simply being aware, whilst an undeniable experience, has no objective qualities. It is a non-objective experience. Being without form, or non-objective, the experience of being aware, or awareness itself, seems like nothing from the point of view of mind, which can only know objective experience, that is, thoughts, feelings, sensations and perceptions. It is therefore usually overlooked or ignored.

The experience of being aware of being aware may be triggered by the thought 'Am I aware?' but it is revealed *between* this thought and the response 'Yes', that is, in between two appearances of mind. Just as the colourless screen is all that is present between two frames of a movie, so the experience of being aware or awareness itself is all that is present between two appearances of mind. However, just as the transparent screen – the background and reality of the movie – does not itself appear as an object in the movie, so the experience of being aware can never be experienced by or appear in the mind, even though it is the background and reality of the mind itself.

Likewise, just as the screen is revealed briefly between two frames in the movie but also remains present *throughout* the movie, although it seems to be obscured by it, so the non-objective experience of being aware is not just present between two thoughts or perceptions but remains present during their existence as their background and reality, although it is usually obscured by them.

Being aware, or awareness itself, is the ever-present, self-aware background upon which all experience appears. Being aware is to experience as a self-aware screen is to a movie. Awareness is the essential, irreducible nature of mind when all objective form or content has been removed from it, just as the screen is the essential, irreducible nature of the movie when all images have disappeared.

All experience is a play of mind, and mind – that is, thinking, feeling, sensing and perceiving – is a self-colouring of awareness, just as the movie is a colouring of the screen. Experience is the modulation or activity of awareness, just as a movie could be said to be the modulation or activity of the screen. When the mind ceases, an entity doesn't come to an end; awareness only ceases to colour itself with its own activity and remains in its essential, uncoloured condition. Likewise, when the activity of mind begins, nothing starts or comes into existence.

Mind is not an *entity* that is distinct from awareness; it has no status or existence of its own. In fact, there is no such thing as 'a mind', just as there is no such thing as 'a movie' with its own existence independent of the screen. Mind is the *movement, activity or functioning* of awareness. Awareness itself, or the non-objective experience of being aware, is the essential, irreducible essence of mind, just as the screen could be said to be the essential, permanent element of the movie. In fact, awareness is not just the background of the mind; it is *all there is to mind*. Mind is a temporary limitation of unlimited awareness, which is itself all that is ever truly present.

The question 'Am I aware?' and the response 'Yes' are thoughts. As such, they are mind, or a self-colouring of awareness. In between these two thoughts awareness does not cease to exist, any more than the screen, relatively speaking, ceases to exist in between two frames of a movie. Its original, essential, irreducible essence is simply revealed. It ceases to veil itself with the activity of mind and stands naked, simply knowing its own unconditioned and essential nature of pure, indivisible, infinite, aware being – pure, indivisible and infinite because there is nothing in itself other than itself with which it could colour, divide or limit itself. Being indivisible, there is only one light of awareness.

In the form of mind, awareness cannot know its own essential, uncoloured condition, just as it is not possible to see a white page through a coloured lens. However, in the pause between two thoughts, awareness 'becomes' aware of itself. It recognises itself, or knows again something that it has always known – or rather eternally knows – but seemed to forget when it coloured itself in the form of objective experience.

Awareness never ceases being its own essential, irreducible 'self'. It only seems to become something else – a finite mind – when it colours itself with the activity of thinking, sensing and perceiving, thereby obscuring its essential, unconditioned nature from itself. All there is between two thoughts or perceptions is the experience of being aware, or awareness itself. Therefore, it is awareness that recognises itself or knows itself again in that timeless pause. There is no other entity present. Awareness's nature is to be aware, and just by being aware it is aware of itself. It is inherently self-aware.

This is why Balyani said, 'I knew my Lord through my Lord.' To suggest that God's infinite being is known by anything other than itself is the ultimate blasphemy. It is to posit a second entity, a self, apart from God's infinite being, a finite consciousness apart from infinite consciousness. As Balyani said, 'No one but He knows Him.'

*   *   *

Awareness's primary and fundamental knowledge is the knowing of its own being, the consciousness of its own existence. It knows itself before it knows any other thing. In fact, awareness cannot *not* be aware of itself, although its awareness of itself is sometimes obscured when it colours itself in the form of mind and, as such, seems to know something other than its own being. All minds arise in and are a self-colouring of the same eternal, infinite light of pure knowing. The 'I' in each of us is the same 'I', modulating itself in and as all apparently separate minds but essentially the same indivisible, self-aware being.

The knowledge 'I am' that shines in each of our minds and that remains present throughout all experience is the same light of pure knowing, refracted into an apparent multiplicity and diversity of minds. Just as the space in all buildings is the same unlimited space, seemingly divided into a multiplicity and diversity of spaces of different shapes and sizes, so the knowing that shines in each of our minds is the *same* knowing, only seemingly divided into a multiplicity and diversity of minds by its reflection in numerous bodies.

Once we have overlooked our unlimited being and, as result, believed and felt ourself to be a temporary, finite awareness, then love, the knowledge of our oneness with all beings, is veiled. It is for this reason that all apparently separate selves long, above all else, for love. Our longing for love comes from the intuition of our shared being. It is the longing that resides in the hearts of all apparently separate selves to be divested of their separateness and returned to their original wholeness or oneness. Love is the experience of that oneness of being. As such, love is God's presence in the heart. That is why most people recognise that love is life's consummate meaning and purpose.

Each of our minds has access to its own infinite reality through the simple experience of being aware or the knowledge 'I am'. The experience of being aware or the knowledge 'I am' is God's signature in the mind. Being aware of being aware is thus the portal, the means and, at the same time, the goal of the mind's quest for absolute truth.

The experience of being aware – the essence of the mind – or the feeling 'I am' – God's presence in our heart – is thus not only the foundation of happiness or fulfilment in individuals, but the ultimate source of peace amongst communities and nations. Thus, in the Direct Path the two ways of knowledge and devotion find their ultimate resolution: they merge and become indistinguishable.

*　　*　　*

Awareness's knowing of its own being is the simplest, most obvious, intimate and ordinary experience, which all people have in common. In fact, it is not an experience that a *person* has. Awareness is the experiencing element in all experience. It is *awareness* that has the experience of being aware, that is, it is *awareness* that knows itself in each one of us. The simple knowing of our own being – its knowing of itself in us – shines in the mind as the knowledge 'I am' or 'I am aware', and in our feelings as the experience of peace, happiness or love.

However, because this knowing of our own being has no objective qualities, it cannot be registered by the object-knowing mind. In fact, the presence of the mind seems to obscure it or prevent it from being known, so it is usually overlooked or ignored.

To say that the mind seems to obscure or prevent the knowing of our own essential, irreducible being is not meant to suggest that mind is an entity in its own right that has an obscuring power. The mind is to awareness what an image is to a screen: simply a modulation of it. It is the screen itself that takes the shape of the image with which it seems to be veiled. Likewise, it is awareness itself that, vibrating within itself, assumes the form of the finite mind, from whose point of view it subsequently seems to be veiled or missing. In other words, awareness seems to obscure itself from itself with its own creativity. Awareness seems to lose itself in the very forms that it assumes.

In our culture we primarily value objects, states of mind, ideas, beliefs, opinions, feelings and relationships. We give our attention to objective knowledge so exclusively that our primary experience – the subjective knowledge of 'I', awareness's knowledge of itself, or the experience of simply being aware – is, in most cases, overlooked or ignored.

How is experience known? With what do we know our experience? How do we know that we exist? Is it not extraordinary that, in spite of the fact that all that is or could ever be known is the *knowing* of experience, that knowing itself, the experience of being aware, is almost completely ignored by our culture and its nature rarely investigated? In order to answer such questions from experience, we have to 'go to' the knowing with which all experience is known; we have to 'go to' that which is aware of our experience. We have to become aware of the experience of being aware. Awareness has to become aware of itself.

In fact, awareness doesn't *become* aware of itself; rather, it ceases assuming the form of the finite mind in which it directs the light of its knowing or attention towards an object or state. As a result, it 'returns to itself' and recognises its own ever-present being. In fact, awareness is always aware of its own being, but this knowledge is temporarily veiled, forgotten or obscured when awareness assumes the form of mind and, as such, directs the focus of its attention towards objects.

When awareness relaxes the focus of its attention from objects, it cannot and need not redirect its attention towards itself, for awareness itself is the *source* of attention. Awareness need only rise in the form of mind or attention in order to know something that is *other* than itself. To know itself, awareness does not need to assume the form of the finite mind or attention and, as such, direct itself towards itself, just as the sun does not need to direct its light in any particular direction in order to illuminate itself.

When awareness withdraws the focus of its attention from objective experience, its knowing begins to sink back into itself, and in doing so it is progressively relieved of the limitations it acquires when assuming the form of the finite mind. When awareness ceases to rise in the form of mind or attention, it is, as it were, unveiled and simply knows or recognises its own being alone. Awareness rises in the form of the mind to know the body and the world, but to know itself it need only rest in and as itself; it need only be itself alone.

*   *   *

Prior to any manifestation, awareness remains motionless and alone, knowing only its own eternal, infinite being. Awareness does not know itself as an object in the way the mind seems to know objects, and thus awareness's knowing of its own eternal, infinite being is said to be 'empty' or 'void'.

However, that is only true from the point of view of the mind, which believes objects to be real things in their own right, made out of stuff called 'matter'. From such a point of view awareness is empty, void, not-a-thing or nothing. From its own point of view – which is the only *real* point of view, and is itself not a 'point' of view – awareness is not nothing, nor is it something. 'Nothing' and 'something' both belong to mind, for both derive their meaning from the assumption of independently existing 'things'.

The finite mind can only know a limited object – something with limited, objective qualities, such as a thought, feeling, sensation or perception – even though it is itself made out of unlimited awareness. The mind cannot even think of awareness, for awareness has no objectives qualities and the mind can only think of something objective. If the mind tries to think of awareness, it will either imagine a blank, empty object or state or, faced with the impossibility of the task, come to an end.

In fact, the mind doesn't come to an end, for the mind is not an entity in its own right that begins and ends. When it is said that the mind comes to an end, it is meant that awareness ceases vibrating within itself and returns to its original, objectless, 'motionless' condition. All that comes to an end when the mind supposedly comes to an end is the *activity* of awareness, which awareness itself freely assumes in order to take the form of the finite mind and, as such, appear to itself as the world.

We cannot even legitimately say that awareness 'returns' to its original condition, any more than we can say that the screen returns to its original condition when the movie ends. Awareness is always in its original condition. It is only from the point of view of the finite mind – the activity that awareness assumes in order to manifest its infinite potential in the form of objective experience – that awareness seems to cease being in its original condition and to become an object, other or world. From its own point of view, it never becomes anything other than itself or ceases to be itself alone.

In relation to the materialist belief that the body and world are made out of something solid, material and 'full', it is legitimate to say that awareness is not material, that it is 'empty' of the solid substance out of which objects are supposedly made. However, that statement starts with the idea of matter and works its way up to awareness, instead of starting with awareness and staying with awareness alone. Awareness is the only legitimate place to start, because awareness is the primary and, in fact, the only element present in experience.

In reality, awareness is neither full nor empty. The idea that awareness is empty is a thorn that is used to remove another thorn: the idea that it is a by-product and shares the limits and destiny of matter. Awareness is beyond or prior to both fullness and emptiness, and cannot be conceptualised by the mind. We are, however, using language that has been designed to describe dualistic experience – that is, the apparent subject–object relationship in which an inside self made of mind supposedly knows an outside object, other or world made of matter – and so we have to make a concession and use such language as skilfully as possible to point towards an experience that ultimately cannot be described.

The irony is that the mind that tries to describe reality is presently creating the very duality from which it is simultaneously trying to emerge. The mind can never find, let alone describe, the reality that it seeks, for it is itself the very activity that seems to divide that reality into a multiplicity and diversity of objects and selves, each with its own name and form that can be described in language.

One might then question the legitimacy of a book such as this, or indeed any attempt to approach and describe the reality of experience, and from an absolute point of view such an objection is reasonable. In fact, all the skilful means prescribed by the religious and spiritual traditions are compassionate concessions to the mind that seeks its own reality, whether that search is felt as the desire for knowledge and understanding, the longing for peace, happiness or love, or devotion to God's infinite being.

The mind that explores these matters is like a moth that seeks a flame. The moth is attracted by the light of the flame, just as the mind is attracted by the gravitational pull of its source and essence. However, as the moth approaches, the flame becomes hotter and hotter and the moth begins to dance around it, at once attracted and repulsed by the heat, in which it intuits it will simultaneously die and discover its heart's desire.

Such is the play of love and resistance with which the separate self seeks and resists its true nature, intuiting that everything for which it truly longs is to be found there and knowing at the same time that in order to experience it, it must die into it. As the Sufi mystic and poet Rumi said, 'In the existence of your love, I become non-existent.'

Likewise is the value of all spiritual discourse and practice that draws the mind inexorably inwards towards its source and reality, until at some point the mind loses its limitations and stands revealed as the very reality for which it was in search.

CHAPTER 8

THE ESSENCE OF MEDITATION

Meditation is not an activity that is undertaken by mind. It is the very opposite of mind's activity. Awareness needs to arise in or assume the form of mind in order to know objective experience, but to know its own being it does not need to assume the form of mind. Indeed, it cannot know its own being in the form of mind.

To know its own being as it is, awareness need only rest in and as itself. That is, it need only *be* itself. But awareness is *already* itself; it doesn't need to go anywhere or do anything special to be and therefore know itself, just as the sun doesn't have to go anywhere or do anything to illuminate itself.

Meditation is not an activity of mind but rather a relaxing, dissolving or sinking of mind into its original, unconditioned, unborn essence. However, mind is not an *entity* that can relax, dissolve or sink into its essence like the sun sinks into the western sky. Mind is the *activity* that awareness itself freely assumes in order to manifest and know its infinite potential. Thus, the only entity – if we can call it an entity – present in mind is awareness itself.

So when it is said that meditation is a relaxing of mind into its essence, what is meant is that in meditation the activity of awareness, namely mind, gradually diminishes. Mind is awareness in motion; awareness is mind at rest. In this relaxation, mind is gradually, in most cases, but occasionally suddenly, divested of its accumulated conditioning, leaving its essential nature of pure awareness exposed, simply knowing its own unlimited being.

Awareness is the experience of simply being aware, or aware being. It is our essential, irreducible, indivisible nature. All objective experience can be removed from awareness, but awareness can never be removed from itself. This does not mean that thoughts, feelings, sensations and perceptions can be removed from awareness in the way that physical objects can, relatively speaking, be removed from the space of a room, although in the early stages of this exploration it is reasonable to say so. That formulation is valid for establishing the presence and primacy of awareness, but once this has been done, like all formulations it must be abandoned.

We never, in fact, know or experience discrete objects called thoughts, feelings, sensations and perceptions that have their own existence independent of awareness or knowing; we only know thinking, feeling, sensing and perceiving. That is, we only know experiencing, and all experiencing is a modulation of the knowing with which it is known and out of which it is made. It is not that thoughts, feelings, sensations and perceptions appear and disappear *in* awareness, like clouds in the sky. They are self-modulations *of* awareness, just as an image is a modulation of the screen but never exists or 'stands out from' the screen with its own independent and separate identity.

Mind is the activity in whose form awareness manifests and knows objective experience. Meditation is a relaxing of that activity, leaving awareness, its essential reality, standing alone in its original, naked, uncoloured, unmodulated condition – pure, luminous, empty knowing. Meditation is to be knowingly that luminous, open, empty, space-like presence of awareness, knowing its own aware being.

*   *   *

Awareness is self-aware. It knows itself simply by being itself. Therefore, no special activity is required by awareness for it to be aware of itself as it is. In fact, a cessation of the mind's activity is required, because the mind's activity is the form in which awareness appears to itself as something *other than* itself, that is, objective experience.

The mind that tries to undertake a special activity called 'meditation' in order to recognise its essential nature, such as focusing its attention on a subtle object or controlling the breath, is like a character in a movie that travels the world in search of the screen. The screen never appears as an object in the movie, although all objects are made only of the screen.

However, as a concession to the character in the movie in search of the screen, we may draw her attention to a subtle object such as the physical space all around her. The empty space that the character experiences is a hint in her three-dimensional world of the reality of her experience, the two-dimensional screen.

We could say that the empty space in which the character's experience seems to appear is the first form of the screen in her world. The two-dimensional screen appears as three-dimensional space in her world due to the limits of her mind, and thus empty physical space is like a place-holder for the presence of the screen. It is still an object in her world, but it is an empty object that mimics the presence of the screen. If the character gives her attention to this empty space, her fixation on objects will relax and her mind will be open and receptive to an intuition of the screen.

The character in the movie can never realise she is the screen, because it is her identity as a limited character that is itself the veiling of the screen. In order to recognise her identity as the screen, the very mind which she is directing towards the empty space must dissolve. In other words, she must cease being the separate character.

In the same way, pure awareness never appears as an experience in or of the mind, although the feeling of being or the knowledge 'I am' is a hint in the mind of its presence. As the mind gives its attention to the feeling of being or the knowledge 'I am', its fixation on objective experience is relaxed and it begins to dissolve into its source and essence.

The feeling of being or the knowledge 'I am' is the last frontier of the mind as it seeks its source and the first frontier through which awareness passes as it manifests objective experience. This is why Jiddu Krishnamurti referred to it as 'the first and last freedom'. Awareness loses its last freedom as it passes through the portal 'I am' and assumes the form of the finite mind, and the finite mind gains its first freedom when it passes through the same portal in the opposite direction on its return to infinite awareness. Thus, in order to recognise its essential reality, mind must cease being mind. Mind cannot know awareness, even though it is made of it. Only awareness can know awareness.

*   *   *

The knowledge 'I am' is the mind's access to the absolute knowledge which lies behind, and is the ultimate reality of, all of its relative knowledge and experience. It is to this quest that Tennyson refers when he speaks of 'yearning in desire to follow knowledge like a sinking star, beyond the utmost bound of human thought'.

The turning of the mind towards its source of pure awareness may be effected by asking a question such as, 'Am I aware?' or 'Who am I?' In order to find the answer to these questions, the mind must look for the experience of simply being aware. As such, the question 'Am I aware?' or 'Who am I?' invites the mind – 'like a sinking star' – away from its customary objects of knowledge and experience – 'the bounds of human thought' – and draws it inwards towards its subjective source, the transparent, luminous, non-objective experience of being aware or pure awareness itself.

However, the mind does not have to go somewhere or indeed do anything to effect this dissolution of itself. There is no distance between mind and awareness, just as there is no distance between the character in the movie and the screen. The character in the movie is only an entity in her own right from her own limited and ultimately illusory perspective. From the perspective of the self-aware screen, the only entity in existence is itself. Likewise, mind is only an independently existing entity from its own limited point of view.

In reality, there is no such thing as 'a mind'. What is normally considered the mind is a temporary, self-assumed limitation and localisation of the indivisible field of infinite awareness itself. From the perspective of awareness, there is only itself and its modifications in the form of mind, but never the absence of itself, nor the presence of any other, independently existing entity. It is for this reason that in the true non-dual teaching, emphasis is placed on recognising the nature of reality rather than on dealing with or trying to get rid of a separate self and its attendant suffering.

So the suggestion to turn around or investigate one's true nature is made as a concession to the mind which feels that it is at a distance from and other than its source of pure awareness. For a mind accustomed to directing the light of its knowing towards objects, the suggestion to turn its knowing upon itself will initially seem to require an effort, just as one who is accustomed to clenching their fist for some time will seem to have to make an effort to open it. Only later will it become apparent that the opening of the fist was not a new effort, but rather the relaxation of a previous effort that had become so habitual that it was no longer noticed as such.

In the same way, only later will it be noticed that to be knowingly the presence of awareness is our natural condition – that is, it is awareness's natural condition – and we cannot therefore make any effort to *be* it. All effort would take us away. In fact, all activities of mind require a more or less subtle effort, an activity of thought or perception. The limited awareness known as mind, and the apparent separate self or ego upon whom it is predicated, is an activity rather than an entity. Meditation is what we are, not we do; the separate self is what we do, not what we are.

In almost all cases the mind does not dissolve in its source immediately; it is a gradual process in which mind directs itself towards the knowing with which it knows its knowledge and experience and, in doing so, sinks more and more deeply into the experience of simply being aware or awareness itself. As Rumi said, 'Flow down and down and down, in ever-widening rings of being.'

As the mind flows down and down, sinking progressively deeper into its own essence – the experience of simply being aware or awareness itself – it is gradually divested of its colouring or conditioning and, in doing so, becomes increasingly transparent and luminous. Pure awareness itself is completely transparent; it has no form, colour or objective qualities and is, as such, without limitation. Pure awareness – the essential, irreducible nature of mind – is, in its own experience of itself, thus eternal and infinite.

As the finite mind sinks into its infinite essence it gradually loses its colouring or conditioning, like an image fading on a screen, and as it does so it loses its limitations. In the non-process of meditation, the mind, as Rumi suggests, becomes progressively 'wider' until it is divested of all its limits and stands revealed as luminous, empty awareness itself, the original, naked, unborn, irreducible and essential condition of the mind, or, in religious language, God's infinite being.

The more interested the mind becomes in its own essential nature of pure, objectless awareness, the more deeply it is drawn into it, and in time this interest grows into an intense love. Nothing could be more interesting or lovable than to know the nature of that through which all things are known. In fact, the mind cannot know what anything truly is until it knows the nature of the knowing with which it knows its knowledge and experience, that is, until it knows the nature of itself.

*   *   *

The question 'Am I aware?' – or any similar question, such as, 'Who or what am I?', 'What is it that knows or is aware of my experience?', 'From where do thoughts, feelings, sensations and perceptions arise?' or 'What element of my experience never disappears?' – is a unique question, because unlike questions that take the mind on a journey of objective or 'outward' exploration, it takes the mind on an objectless journey in which the knowing with which the mind usually knows objective experience is drawn 'inwards' or 'selfwards' towards its own essential, uncoloured reality.

As the mind travels inwards, its essential quality of pure knowing gradually, in most cases, loses its colouring. Just as the slowly fading image seems to reveal the screen, which was in fact always in plain view, so the mind's essential nature ceases to obscure itself in the form of objective experience and is revealed to itself as pure, objectless, infinite awareness. The mind recognises its own nature: original mind, pure consciousness, infinite awareness.

The question 'Am I aware?' or 'Who am I?' initiates access to the highest intelligence of which the mind is capable. It is the ultimate thought. The mind can go no further than that; it is the farthest shore of knowledge. It is for this reason that Shantananda Saraswati, the former Shankaracharya of the north of India, said, 'Real thinkers don't think.' That is, a mind which pursues absolute truth or a heart that longs for unconditional love will eventually bring itself to its own end.

What it is that triggers this sacred question in each of our minds varies from individual to individual, but sooner or later the quest for knowledge or love must lead to this question. 'Am I aware?', 'Who am I?', 'What is the nature of the knowing with which experience is known?', 'From what does the mind arise?', 'How do I know that I am aware?' – any such question turns the mind upon itself. These are all variations of the same sacred question, and it is in the form of this question that the divergent disciplines of science and religion are united. The desire for knowledge and the love of God are realised to be the same quest.

This sacred question triggers a process in the mind whose resolution is the foundation of all true knowledge and love. It is thus at once the ultimate science, the essence of meditation and the most sacred prayer. It is for this reason that Rumi said, 'I searched for myself and found only God; I searched for God and found only myself.'

Any knowledge that is not founded upon the mind's recognition of its own essence of infinite, indivisible awareness will inevitably share the conditioning and thus the limitations of the finite mind with which it is known, and will therefore always be subject to change and doubt. As such, it will be, at best, only relatively true. Awareness's knowing of its own eternal, infinite being is the only absolute knowledge, for it is the only knowledge that is not relative to and limited by the conditioning of the finite mind. It is the only knowledge that is absolutely true at all times, in all places, for all people and under all circumstances.

A truly civilised culture is one in which all branches of knowledge – politics, psychology, medicine, science, sociology, economics, philosophy, the arts and religion – are founded upon the recognition of the eternal, infinite nature of awareness, the ultimate reality of all experience that knows itself alike in each one of us as 'I' or 'I am', irrespective of nationality, age, gender, race, creed, education, health or wealth. In such a culture each branch of knowledge would tailor the absolute truth to the various fields in which it operated, bringing to humanity its creative, healing intelligence and love.

If there were ever to be a philosophy, religion or science that could truly unify the human race and bring lasting peace, justice and equality to individuals, families, communities and nations, it would have to be based on the one experience that all beings share in equal measure and to which all beings have equal and unlimited access at all times: the knowing of our own infinite being, which shines in each of our minds as the knowledge 'I am'. The fact that all people refer to themselves by the same name – 'I' – is a hint in common parlance of the understanding that we all share the same being.

The experience of being aware is the most fundamental, ordinary, familiar and intimate experience. It shines in all minds as the knowledge 'I'. As such, the 'I' of the finite mind is God's infinite 'I', the only 'I' there is; the indivisible, aware being from whom all finite minds derive their essential identity; the single reality that each mind modulates in a unique way, partially revealing the splendour and beauty that lie at its source. To be that knowingly is the essence of meditation and the art of life.

CHAPTER 9

THE OUTWARD-FACING PATH: COLLAPSING THE
DISTINCTION BETWEEN CONSCIOUSNESS AND OBJECTS

The spiritual path could be divided into three steps. The first step involves the investigation into the essential nature of the ego or separate self through the *neti neti* process, in which the witnessing subject of experience is extricated from all objective content and stands alone as pure consciousness, the primary and fundamental element of all experience.

In the second step, consciousness releases its attention from the objective content of experience, from which it separated itself in the first step, and begins to flow backwards or inwards into itself, eventually coming to rest in itself. It is in this self-resting or self-abiding that consciousness is gradually, in most cases, divested of its self-assumed limitations and recognises its own ever-present and unlimited being. This self-resting or self-abiding is the essence of meditation and prayer.

Once consciousness has recognised its own ever-present and unlimited nature – the recognition that is traditionally referred to as enlightenment or awakening – the purpose of distinguishing consciousness from objects has been accomplished and it is now necessary to dissolve this distinction. Thus, the third step on the spiritual path involves an exploration of objective experience in the light of our new understanding in order to collapse the apparent distinction between consciousness and its objects.

In this exploration, we discover that consciousness is not simply the witnessing presence *to* which all experience appears but the space or field *in* which all experience appears. Try to find an experience that takes place outside consciousness. All that is known of the apparently outside world is perception – sights, sounds, tastes, textures and smells – and all perception takes place in consciousness.

Even if we perceive something that seems to be at a vast distance from ourself, such as the moon, all that could ever be known of it is a thought, image or perception, and all thoughts, images and perceptions appear in consciousness. Likewise, all that is or could ever be known of the body are sensations and perceptions, and all sensations and perceptions appear in consciousness. Thus, if we stay close to the facts of experience, the world and the body are appearances *in* consciousness. Consciousness can never know or come in contact with anything outside of itself.

Try now with your attention to leave the field of consciousness in which all experience appears, in the same way that a child might lie in bed wondering how far space goes and what, if anything, might lie beyond it. See that attention never leaves the field of consciousness. All experience takes place in and is known by consciousness, and as experience is all that is or could ever be known, we cannot legitimately claim the existence of anything outside of consciousness. To do so would require a leap of faith.

Thus, consciousness itself could be likened to an open, empty field or space-like presence in which all objective experience appears, like clouds appearing in an empty sky. Consciousness is not, in fact, a space; it is dimensionless. However, it is not possible to think of or visualise something without dimensions, so as a concession to the mind that wishes to think and speak of the nature of reality, it is legitimate to add a space-like quality to consciousness and to describe it as an open, empty, space-like field or presence.

*   *   *

Notice that, just as no cloud is at a distance from the sky in which it appears, no experience is at a distance from the consciousness in which it arises. This recognition collapses, at least to a degree, the apparent distance between consciousness and its objects, although a distinction still remains, just as there is a distinction between the sky and the clouds.

Does a cloud ever have its own existence independent of the sky? Can a cloud be removed from the sky and still maintain its existence? Is not the temporary existence of the cloud borrowed from the permanent reality, relatively speaking, of the sky? Is it possible to have an experience that is independent of consciousness?

What is the relationship between the known object and the knowing field of consciousness in which it appears and with which it is known? Is there indeed anything to the known other than the knowing of it? Investigate your experience – your current, remembered or imagined experience – and ask yourself if you ever find or come in contact with anything other than the *knowing* of it.

All there is to experience is thinking, imagining, feeling, sensing and perceiving, and all thinking, imagining, feeling, sensing and perceiving are intimately one with the consciousness with which they are known and in which they appear. That is, all there is to thinking, imagining, feeling, sensing and perceiving is the knowing of them. There is not one thing called 'thinking', 'imagining', 'feeling', and so on, and another called 'the knowing of it'.

Thinking *is* only the knowing of it. Feeling *is* only the knowing of it. Perceiving *is* only the knowing of it. All there is to experience is the knowing of it. In fact, we never come in contact with the 'it'. The 'it', the object of experience, is never found to exist independently, just as a cloud is never found to exist independently of the sky.

Whilst this is a legitimate stage of understanding, and the metaphor of clouds in the sky serves to evoke the provisional relationship between consciousness and its objects, it is still a position of duality; we could call it enlightened duality. No object or thing ever 'ex-ists' or stands out from consciousness in the way that a cloud seems to stand out from the sky as an independent object in its own right. So we must now abandon this metaphor and replace it with one that more accurately reflects our experience. Thus, consciousness is to experience, not as objects are to a witness, nor even as clouds are to the sky, but rather as a self-aware screen is to the movie that is playing upon it.

Even this metaphor suggests that there is something called 'a movie', albeit dependent for its existence upon the screen, but that is not true. We do not even find an object called a movie that is *dependent* on the screen, let alone independent of it! All there is to a so-called movie is the reality, relatively speaking, of the screen. The movie doesn't exist or stand out from the screen. There is no such thing as 'a movie' as an object in its own right. All that is ever found is the screen, modulating itself in the form of the movie but never ceasing to be the screen and never becoming anything other than itself.

Likewise, we never know or come in contact with an object of experience, be that object a thought, feeling, sensation or perception. We only know knowing. Knowing is all that is or could ever be known. And what is it that knows this knowing? Only that which knows could know knowing. Only knowing could know knowing. The knowing of experience, or the consciousness of experience, is all there is to experience, and it is knowing that knows this knowing: knowing, knowing only knowing.

In the final stage of this exploration the distinction between consciousness and its objects collapses completely. Experience is not just known *by* consciousness; it does not just appear *in* consciousness; *consciousness is all there is to experience*. There is only consciousness. As the Vedantins say, 'There is only the Self', and as the Sufis say, 'Everything is God's face'.

*   *   *

There is no part of experience that is not made of knowing, and thus all that is present in experience is knowing or consciousness itself. Indeed, there are no *parts* to experience, period. All there is to experience is knowing or consciousness, and there is nothing in consciousness other than consciousness itself which could divide itself into a multiplicity and diversity of objects, parts or selves. Experience is a single, indivisible whole.

In consciousness's experience – and consciousness is the only 'one' that has or knows experience – there is only itself, either at rest knowing its own infinite, inherently peaceful and unconditionally fulfilled being, or in activity knowing itself, in the form of mind, as the world. As such, all experience is infinite, indivisible consciousness assuming the activity of mind and appearing to itself as a multiplicity and diversity of objects and selves, but never actually being, becoming or knowing anything other than itself alone. There is only God's infinite being.

Consciousness is the self-aware screen that is modulating itself in and simultaneously knowing itself as all forms of experience. When consciousness is modulating itself in the form of experience, it is known as mind; when it is not modulating itself it is known as pure, unconditioned and, therefore, unlimited consciousness. When the screen is modulating itself it is known as a movie, and when it is not doing so it remains the 'pure', uncoloured screen.

The finite mind, separate self or ego is the agency through and as which consciousness simultaneously assumes the form of and knows manifestation, just as a movie could be said to be the activity through and as which a screen appears in the form of a landscape. However, just as the screen modulates itself in the form of the movie but never ceases to be the screen, so consciousness never actually becomes the finite mind, separate self or ego.

It is for this reason that in the non-dual traditions in general, and in the Advaita Vedanta tradition in particular, the finite mind, separate self or ego is said to be an illusion. This does not mean that what is conventionally considered to be the finite mind is simply not there, as is believed in many contemporary, so-called non-dual teachings, but rather that what *is* there is not what it seems to be. An illusion always has a reality to it. A mirage in a desert is an illusion as water, but light, relatively speaking, is its reality. Likewise, the finite mind is an illusion as a separately existing entity, but infinite consciousness is its reality. The apparent 'I' of the separate self is the true and only 'I' of eternal, infinite consciousness.

It is only from the limited and ultimately illusory perspective of the separate subject of experience that there are a multiplicity and diversity of separate objects. As such, the apparent subject and object of experience are two sides of the same coin, the infinite, indivisible reality of consciousness. Thus, we could define consciousness as that with which all experience is known, in which all experience appears and out of which all experience is made.

As that *to* which all experience appears, consciousness is the separate witness of experience. As that *in* which all experience appears, consciousness is an open, empty, space-like presence in which objects appear like clouds in an empty sky. As that *out of* which experience is made, consciousness is the dimensionless, self-aware screen which is itself simultaneously assuming the form of experience and knowing it.

From this perspective, perception *is* creation. The act of perception itself brings creation out of potential in infinite consciousness and into existence. But even when in existence, there is nothing truly present in creation other than infinite consciousness itself.

Thus the final stages of the spiritual process involve a return to objective experience and the dissolution of any apparent distinction between consciousness and its objects. This recognition brings enlightened understanding into the realm of our activities and relationships. It is the fruit of enlightenment, in which the peace, happiness, freedom and love that are inherent in consciousness's knowing of its own infinite being gradually infiltrate, permeate and saturate all realms of objective experience, thereby progressively colonising and outshining it in the light of pure knowing.

*   *   *

The consciousness of experience is all that we are ever conscious of, and the 'we' that is conscious of it is consciousness itself. In other words, consciousness is the only substance present in experience. If experience is all that is or could ever be known, the only statement that we can make with absolute certainty is that there is only consciousness, and that consciousness, being the totality, is eternal and infinite, for there is nothing in itself, other than itself, with which it could be limited. In other words, the knowledge that only God's infinite, self-aware being is all there truly *is* is not simply a religious belief; it is the ultimate knowledge, the only absolute knowledge there is.

Even that is not quite right, for to assert that consciousness is infinite, that is, not finite, is to imply that there are finite things that consciousness is not. It is to define what *is* in terms of what is *not*. In fact, all words describe objective experience, and once we have discovered that there are no objects, either dependent or independent, it becomes clear that all words describe what is not. Thus it is not possible to say a true word about what *is* or what is *real*. It is not possible to say a true word about consciousness, our self or God's being.

Even to name 'that which is' as consciousness, being, God or self is to say too much, for each of these words derives its meaning by comparison with its opposite. It is for this reason that the ancient sages, in their wisdom and humility, preferred only to say what reality is not, rather than making any positive assertions about it. Even to say that reality is one would be to make too positive an assertion, and hence they said that it is simply 'not two', in Sanskrit *ad-vaita*. However, this should not be confused – as it often is in New Age non-duality and Neo-Advaita circles – with the post-modern belief that in the absence of any absolute truth, everything is relative.

Some sages simply remain silent, preferring not to restrict absolute reality within the confines of the mind, whilst others use the mind as best they can to evoke this recognition, using such terms as infinite, immortal, impeccable, indestructible, imperturbable, irreducible, unconditioned, uncreated, unborn, undefiled, unharmable, unaffected and untouched. Although these seem to be adjectives that describe what consciousness or reality *is*, they are all, in fact, negative statements indicating what it is *not*.

When we have denied to consciousness or reality all the qualities that are usually ascribed to objects, it stands revealed, shining in and as itself, and all experience is seen to be only that. The beautiful and deceptively simple statement in the Upanishads, 'I am That', when properly understood conveys this insight and is, as such, a condensed expression of the highest understanding, which has the power to awaken a sensitive and receptive mind to its essential reality.

*　　*　　*

These stages of development, from the conventional position of the ego, through the enlightened understanding of our eternal, infinite nature of pure consciousness, to the establishment of this understanding in all realms of our lives, are detailed in all the great spiritual and religious traditions. From the conventional perspective of a 'mature' adult or ego to the recognition of our essential self as pure consciousness is a path from the belief 'I am something' – a body-mind – to the understanding 'I am nothing', not a thing. The path from 'I am not a thing; I am pure consciousness alone' to the recognition of God's infinite being, or infinite consciousness itself, as the ultimate reality of all things is a path from 'I am nothing' to 'I am everything'.

The Zen tradition describes these three stages in this way: 'First the rivers are rivers and the mountains are mountains. Then the rivers are no longer rivers and the mountains are no longer mountains. Then the rivers are rivers and the mountains are mountains again.' Just as the pre-egoic experience of an animal or infant looks similar to and shares some of the qualities of the post-egoic recognition, so rivers and mountains look similar from both the pre-egoic and post-egoic perspectives, only in the former they are made out of something other than ourself, namely matter, and in the latter their fundamental essence is recognised to be identical to our self, namely consciousness.

The first path, from 'I am something' to 'I am nothing', is an inward- or selfward-facing path of discrimination, which is most clearly elaborated in the Vedantic tradition. The path from 'I am nothing' to 'I am everything' is an outward- or object-facing path which is more clearly elaborated in the Tantric traditions, especially that of Kashmir Shaivism. If the Vedantic tradition is one of exclusion or discrimination, in which the ultimate reality is extricated from the obscuring veil of objective experience, the Tantric tradition is one of inclusion, in which the apparent distinction or separation between consciousness and all objects and others is dissolved. This dissolution of separation is known at the human level as love in relation to others and beauty in relation to objects.

Ramana Maharshi described the same stages of development: 'The world is unreal; only consciousness is real; consciousness is the world', that is, consciousness merged in experience; consciousness extricated from experience; experience merged in consciousness. All the infant knows is experience; all the sage or the mystic knows is consciousness; and all objects, others and the world are only that.

William Blake was trying to explain this to one of his materialist friends, who asked him, 'When the sun rises, do you not see a round disc of fire somewhat like a guinea?' Blake replied, 'O no, no, I see an innumerable company of the heavenly host crying, "Holy, Holy, Holy is the Lord God Almighty!"'*

The same understanding expressed in the Upanishads as 'I am That' is, in the Christian tradition, 'I and my Father are one'. That is, what I essentially am – pure consciousness – and what the ultimate reality of the universe really is, are one and the same. The 'amness' that all beings feel at the core of themselves and the 'isness' or existence that is shared by all objects are modulations of the same indivisible, self-aware being. Being indivisible, it is whole, complete, needing nothing, desiring nothing, fearing nothing and disturbed by nothing, for there is nothing in itself other than itself with which it could be enhanced, diminished or disturbed. Thus, it is peace and fulfilment itself.

The great equation of the Vedantic tradition states simply, '*sat* (being) *chit* (consciousness) *ananda* (happiness)', indicating that when the being or existence that is common to all objects is revealed to be identical to the consciousness with which all experience is known, the distinction or separation between people, animals and the world dissolves, and the experience of peace, happiness, love and beauty shines.

---

*From 'A Vision of the Last Judgment' (c. 1810).

There is only infinite, indivisible consciousness modulating itself in and as the totality of experience but never being, becoming or knowing anything other than itself and, therefore, free of any impulse to fulfil or protect itself. It is, as such, unconditional happiness itself.

This recognition is said to be absolute because it is not derived from or dependent upon anything other than itself. It never changes; it doesn't come and go; it is not relative to the finite mind; and it is known by itself alone. This understanding is the essential recognition that lies at the heart of all the great religious and spiritual traditions. The differences lie only in the forms in which it is expressed and the means by which it may be recognised. It is for this reason that, whilst the truth can never divide people, religion almost always does.

In fact, this recognition is the only understanding that can truly unite people, because it is shared by all people equally, irrespective of race, creed, religion, health or wealth. It remains the same throughout all states and circumstances, and throughout all ages. It can never be the property of any single person, nor the provenance of any single nation or religion. It is equally available to all people and no one has privileged access to it. It does not have to be earned because it is already and always our essential, unconditioned nature. It cannot be possessed or manipulated. It shines in each of our minds as the knowledge 'I am', the feeling of being or the experience of love. It is, therefore, the foundation of peace in individuals, communities and nations. It need only be recognised and its implications lived in all realms of experience.

CHAPTER 10

The objective world as it is known from the perspective of the separate self or finite mind is an illusion, but all illusions have a reality to them. In the words of Huang Po, 'People neglect the reality of the illusory world.' All that is known of the world is the finite mind; the reality of the finite mind is infinite awareness; and nothing ever happens to awareness. Thus it is said that in the ultimate analysis, nothing ever happens.

No finite object or temporary self ever comes into existence, lasts in time, is moved or changed, grows old, dies or disappears. All apparent objects and selves shine with their reality of eternal, infinite awareness alone. Every moment of experience, irrespective of its content, is infinite, indivisible, utterly intimate awareness, our very own self or being, appearing to itself, in itself, as itself.

All objects are said to exist: a chair exists, a thought exists, an emotion exists. As such, existence is the common element, the shared essence, of all objects. However, existence is not a property of objects, just as the screen is not a property of a movie, nor water a property of a wave. Objects do not *have* existence. Existence – the isness of all seeming things – belongs to awareness alone, the only 'one' that truly is.

Even from the conventional perspective, in which mind is considered to be derived from the body and the body from the world, the essential nature of mind must be identical to the essential nature of the world, just as the essential nature of any part must be identical to the whole from which it is derived. Therefore, the essential nature of our self – the amness that shines as the knowledge 'I am', the experience of being aware or the feeling of being – is identical to the isness of all seeming things, the essential nature of the world. Existence is identical to awareness.

This is the radical understanding that is the non-dual essence of all the great religious, spiritual and philosophical traditions. Existence is identical to awareness, for awareness obviously *is* – I *am* – and the isness of awareness must be the same as the isness of existence. If they were different they would each have to have limiting qualities to distinguish them from each other, and if isness had any limitations it would not be the common element in *all* apparent objects. It is only from the illusory perspective of a separate self that existence and awareness are divided into an outside world and an inside self.

No object, thing or separate self truly exists, be it made of matter or mind. Objects and selves borrow their apparent existence from the sole reality of pure awareness. Awareness alone is all that truly *is*. I alone am, and everything and everyone is that! In fact, there are no things or selves for awareness to be the totality of. I am, and all seeming things and selves are that alone; not I, the separate body or mind – there *is* no separate body or mind – but I, the only 'I' there is, infinite, indivisible awareness, God's infinite, self-aware being. Existence *is* awareness itself.

Nothing – that is, no thing – exists! The Latin origin of the word 'exist', *exsistere*, meaning 'to stand out', implies that objects and selves 'come into existence' or 'stand out from' the background of pure being or awareness, in the same way that an image appears to stand out from a screen. But no object or self ever comes into or goes out of existence. Indeed, even when awareness, vibrating within itself, assumes the form of the finite mind and seems to become a multiplicity and diversity of objects and others, known from the perspective of an inside self, no object or self ever actually 'stands out from' itself. No thing ever comes into or disappears out of existence.

Existence belongs to pure awareness alone. In fact, even awareness does not *exist*; it does not 'stand out' from itself. It *is*. I *am*. Awareness cannot 'come out of' itself, because there is, in its own experience of itself, nowhere for it to go other than itself, no dimension into which it could venture or extend itself.

Existence is the activity of being; being is existence at rest. The existence that an object or self seems to possess is appropriated from awareness's being, just as an image borrows its reality from the relative reality of the screen. All that is real in the image is the reality of the screen. All that is real in experience is infinite, indivisible awareness. That means that the entire universe is our self.

How would it be to live all aspects of one's life in a way that is consistent with this understanding? How would we treat animals, people and the earth? This understanding is the foundation of all morality and ethics. When St. Augustine was asked how one should behave in the world, he is said to have responded, 'Love, and do whatever you want!' If we understand and feel that every animal, person and object is our very own self, we cannot go wrong. That is the experience of love. It is the only moral guidance we need.

*　*　*

There is only awareness, God's infinite, indivisible being, which shines in each of our minds as the knowledge 'I am', the feeling of being or the experience of love or beauty. As such, 'I am' is the first form of knowledge. It is the portal through which the world proceeds out of potential in awareness into actuality in finite existence, and the same portal through which the world passes in the other direction as it dissolves into its essence.

The experience of beauty is the experience of the world dissolving into its infinite essence. It is a revelation of infinity. That is why we love to walk in nature, listen to music and experience art. Indeed, it is why we have art in our culture at all.

All existence lies enfolded in potential within awareness itself. Awareness births existence out of its own being by freely assuming the form of perception, thereby collapsing its own infinite potential into a specific form and excluding in that moment all other possibilities within itself. Perception is creation!

This is what William Blake meant when he said, 'Every bird that cuts the airy way is an immense world of delight, closed by your senses five.'* Awareness itself is an immense world of delight. It is peace and joy itself. It is awareness itself that assumes the activity of the five sense perceptions, thereby enclosing its own joyous immensity in form and appearing to itself as the world. The world is a materialisation of that joy.

*From *The Marriage of Heaven and Hell.*

That is the world that an artist wants us to see. A true work of art is a condensation of beauty. It is an invitation to see – no, to *be* – the world. A work of art contains within it the power to effect the dissolution of everything that separates ourself from the object, other or world, and to reveal our self as the immensity that is everything.

We might then ask, why is there a world? The 'Why?' question that has plagued philosophers for centuries can never be answered on the terms in which it is asked. The 'Why?' question is asked by the mind that believes that its dualistic way of perceiving reality is accurate. It is a question that can only be answered by undermining that assumption at its origin. Without the assumption, it cannot stand. To ask 'Why is there a world?' is like asking, 'Why is the earth flat?' Only because we believe it to be so! Thought abstracts the world!

The question 'Why is there a world?' should be reformulated in the light of this understanding as, 'Why is there a finite mind?' There isn't! The finite mind is only a finite mind from the illusory perspective of a finite mind.

The apparent division of awareness into a self on the inside made of mind and a world on the outside made of matter never actually happens, just as a screen is never divided into parts when the movie begins. It is only from the perspective of one of the characters in the movie that the apparent world she sees around herself comprises a multiplicity and diversity of objects whose reality seems to differ from her own. Likewise, it is only from the perspective of the ego or separate self around whom the finite mind revolves that experience comprises a multiplicity and diversity of objects and minds.

In other words, it is only from the illusory perspective of the ego or separate self that duality – the division of experience into mind and matter – actually happens. From awareness's own point of view, there is always only its own eternal, infinite, indivisible reality, shining in and as the totality of experience. There is only eternal awareness or God's infinite being, modulating itself in the form of the finite mind but never ceasing to be or know itself alone.

Why is there duality? There isn't! What is the purpose of apparent duality? To make us realise that!

* * *

Unlike a single mind, which can entertain only one thought or perception at a time, awareness can assume the activity of an apparent multiplicity and diversity of minds simultaneously. Imagine the single, indivisible field of awareness vibrating within itself, and visualise each of our minds, not as an entity in its own right, but as an *opening* onto that vibrating field of experience. Each of our minds is like an opening through which infinite awareness knows itself in the form of the world.

At no point does an actual mind come into existence, let alone a multiplicity and diversity of minds. In other words, each of our minds is a temporary limitation of the true and only mind of infinite awareness, just as the space in every building in the world is a temporary limitation of the single, indivisible space of the universe.

To speak of an individual mind, let alone numerous minds, is a concession to thought. There are no separate, discrete minds. Separate minds are only such from the provisional and ultimately illusory point of view of an apparently separate mind. Each apparently separate mind is a vibrating field in and of the only mind there truly is: infinite, indivisible awareness. Each vibrating field is a borderless flux of loosely gathered energies that is in contact with all the other vibrating fields within the larger field of 'infinite mind' or awareness itself.

However, as a concession to our desire to think about these matters, we may say that awareness vibrates within itself and appears to itself in the form of the world, which it knows from the perspective of each one of our minds. That is, awareness gives birth to itself in the form of the world and simultaneously collapses into a finite mind or separate self within that world, from whose perspective it seems to be seen or known, just as at night the mind dreams the world within itself and simultaneously collapses into a character in the dream, from whose perspective the dream world is experienced.

This subject–object relationship is the agency of manifestation through which awareness actualises its infinite potential. And just as the mind folds the dreamed world up within itself at the end of each dream, so awareness folds the world up within itself at the end of every perception. Of course, awareness does not do this *in time*. Time only comes into apparent existence when awareness assumes the activity of thought. In reality, awareness is birthing and dissolving the world within itself in its own dimensionless presence, the eternal now.

The Big Bang did not take place at a moment of time. There is no time at which or in which the Big Bang, or indeed any event, can take place. Time is the eternal now seen through the prism of the mind's activity. The Big Bang takes place every time awareness collapses its own dimensionless presence and assumes the form of the finite mind. From the point of view of each apparent mind, the world seems to exist outside of and prior to itself, and the Big Bang is considered the moment in time at which the world came into existence. However, from the point of view of reality, nothing ever comes in and out of existence.

There is just awareness itself, being modulated in all forms of experience but never ceasing to be or undergoing any essential modification. Our different experiences are different views of the same reality, each difference due only to the limitations of the mind through which it is viewed. As it says in the Bhagavad Gita, 'That which is never ceases to be; that which is not never comes into existence.'

* * *

Awareness is not the *background* or *knower* of experience, although it is legitimate to say so during the early stages of our exploration into the nature of reality. Awareness is *all there is* to experience, or rather, experience is awareness. There is just infinite awareness, being only itself, knowing only itself, and, because of the absolute non-existence of any distance, separation or otherness in its own experience of itself, loving only itself.

When the Sufis say *La ilaha illa la* – 'There is no God but God' – they do not mean that their God, Allah, is the only true God as opposed to all the other religions' Gods, as is commonly supposed. Rather, they mean that no mind, person, self, object or world ever actually comes into existence. No thing is a thing unto itself. No thing has its own being. The apparent existence of all objects and selves is borrowed from God's infinite, self-aware being, infinite awareness, our very own intimate, impersonal self, from whose point of view there is nothing other than itself. That being shines in the mind as the knowledge 'I am' and in the world as the experience 'it is'. The amness of the self is the isness of things.

The mystic explores the amness of the self; the scientist and artist explore the isness of things. To begin with, these two avenues of research seem to take the mind in opposite directions: the first seemingly inwards into an exploration of the nature of awareness, the second seemingly outwards into an exploration of the nature of existence. However, if both parties are courageous and honest enough, and refuse to stop short of the absolute truth of their experience, they will inevitably arrive at the same conclusion.

That conclusion may or may not be formulated by the finite mind as a series of concepts. It may just as well take the form of a piece of music, a painting, a dance, a poem, an act of kindness or simply a smile at a stranger (who in that moment ceases to be a stranger). If it is formulated in words, as in this book, its formulation will only ever be a pale approximation of the discovery to which it refers. But the discovery itself is not an experience *of* the finite mind or *in* the finite mind, nor is it known *by* the finite mind. That discovery is awareness shining in itself, as itself, by itself, to itself, modulating itself in all forms of experience, but never ceasing to be, know or love itself alone.

CHAPTER 11

THE WHITE RADIANCE OF ETERNITY

In *Adonaïs: An Elegy on the Death of John Keats*, the poet Percy Bysshe Shelley wrote, 'Life, like a dome of many-coloured glass, stains the white radiance of eternity.' The word 'stains' in this context means colours rather than blemishes, borrowing its imagery from stained glass windows. Just as a stained glass window colours the light passing through it, revealing the potential inherent in light itself, so experience or the finite mind refracts the light of pure knowing into the apparent multiplicity and diversity of objective experience, thus bringing part of its infinite potential into finite existence.

Of course, the finite mind is not distinct from awareness in the way that stained glass is distinct from the light that shines through it. The finite mind is the prismic activity of awareness itself, through which its infinite, indivisible and unperceivable nature is refracted into a multiplicity and diversity of objects that are known from the perspective of an apparent subject.

In reality, no experience stains or tarnishes awareness. It is only from the limited and ultimately illusory perspective of a separate self, the protagonist of the finite mind from whose perspective experience seems to be divided into two essential ingredients – mind and matter – that life seems to tarnish or obscure its reality of infinite, indivisible awareness. From the point of view of awareness itself – which is the only *real* point of view because awareness is the only 'one' that knows experience – experience does not stain its own reality, any more than a movie tarnishes the screen on which it appears.

Having said that, awareness cannot really be said to have a point of view; a point of view is precisely what the separate self or finite mind is, that is, a location from which objective experience seems to be known. In order to know objective experience, unlocated, dimensionless awareness must assume a location, place or 'point' from which it can view, know or perceive objective experience, and the body is that location. However, the body is not an object; it is an appearance in the mind, that is, it is the mechanism or agency through and as which awareness locates and thus limits itself, thereby seeming to become a separate subject of experience from whose point of view it can know or perceive objective experience.

This awareness-in-the-body entity is known as 'mind', the seemingly separate subject of experience which borrows its knowing quality from pure awareness and its apparent limitations from the body. However, the knowing with which the awareness-in-the-body entity knows or perceives its experience is not itself located *in the body*, just as the screen is not located in the character in the movie from whose point of view the landscape is viewed. The body is simply the activity that awareness assumes in order to collapse its infinite potential, thereby limiting itself in the form of mind, and thus bring manifestation out of infinite being and into finite existence.

*   *   *

It is easy to check this in our experience. No one has ever experienced, or could ever experience, a static, discrete object called 'a body'. In our actual experience, the body is a flow of sensations and perceptions. And even that is not right; a discrete sensation or perception is never found. We can never freeze a sensation or perception and extract it from the indivisible unity of experiencing or knowing in the same way that we seem, relatively speaking, to be able to isolate an object in space.

All there is to the experience of the body is sensation and perception; all there is to a sensation or perception is the experience of sensing or perceiving; and the only substance present in sensing and perceiving is knowing or awareness. That is, in our actual experience, the body is an appearance in and of the mind, and all there is to mind is knowing. It is thought alone that abstracts a solid, discrete object called 'a body' that it considers separate from knowing.

Having abstracted a discrete object from the indivisible intimacy of pure knowing, thought has to name the substance out of which it is made.

As such, matter could be defined as the material out of which everything outside awareness is supposedly made. The fact that nothing is or could ever be known outside awareness is not deemed sufficient evidence to dissuade most people from believing that there is such a substance.

Exactly the same is true of a physical object or the world. All that is or could ever be known of an object or the world is perception, that is, sights, sounds, tastes, textures and smells. All that is known of sights, sounds, tastes, textures and smells is seeing, hearing, tasting, touching and smelling. And all that is known of seeing, hearing, tasting, touching and smelling is knowing. Thus, if we proceed like honest scientists, allowing reason to be guided only by observable experience, we come to the conclusion that we never experience an object or world as it is conceived by thought.

We do not have to rely on the evidence of science to know the nature of the world. We need only rely on experience, the sole legitimate arbiter of truth or reality. A world made of matter is an abstraction, and it is this calamitous assumption that lies at the root of our materialistic culture and is responsible for separating all objects and others from the intimacy of ourself, thus enabling us to disrespect and degrade our environment and to treat other people and animals in unkind and unjust ways.

The belief in an outside world made of matter is the inevitable corollary of the belief in an inside self made of mind. The two arise together as two sides of the same belief. It is for this reason that Ramana Maharshi said, 'The "I" thought is the mother of the world.' He did not mean literally the thought 'I', but rather that the finite mind is the mother of the world. The first knowledge that appears in the finite mind is the knowledge of its own being – which is formulated as the thought 'I am' – and hence he said that the 'I' thought creates the world.

The same understanding is expressed in the Kashmir Shaivite tradition, which describes the world as an expansion of 'I', and in the Sufi tradition in Rumi's statement that 'Knowledge of the world is a kind of ignorance'. This is not a life-denying statement; on the contrary, it is truly life-affirming. Rumi is saying that conceiving of the world as made out of dead, inert stuff called matter, separate and distinct from ourself, is a denial of the real world; the world as it is actually experienced is made only of the infinite, indivisible knowing which is our own self, the very self which, in religious language, is the self of God's infinite being.

It is this luminous, empty knowing which freely assumes the 'colours' of experience by modulating itself in all forms of thought and perception, but no colour leaves a trace or residue on its essential nature. Life or experience itself doesn't *stain* the white radiance of eternity. Experience is a temporary modulation of that radiance that always leaves it in its original, pristine condition. No experience leaves a trace on awareness. Awareness cannot be harmed, moved, modified, aged, hurt, diminished, enhanced or destroyed by any experience. So we could rephrase Shelley's line as, 'Life, like a dome of many-coloured glass, *colours* the white radiance of eternity': life, in the form of a multiplicity and diversity of objective experience, colours the light of pure knowing.

That knowing – awareness itself, God's infinite being – which shines in each of our minds as the knowledge 'I am', is the common element of all experience, as light is the common element of all colour, and thus does not share the limitations of any particular experience. It is, as such, unlimited or infinite. It does not share the destiny of any of its modifications and is thus ever-present or eternal.

It is only from the perspective of awareness's own self-modification in the form of the finite mind that awareness is temporary and finite. From that limited perspective, objects and selves seem to have an independent existence of their own that starts at one moment in time and space, is subject to change, growth and decay, and ends at another moment. From its own perspective awareness never appears or disappears, nor is it ever moved or changed. It is the eternal, infinite reality of all experience.

In fact, we can go further: the only substance present in the ever-changing appearances of mind is the changeless presence of luminous, empty knowing or awareness itself. Awareness is all there is to experience. It is this luminous, empty knowing which, vibrating within itself, shines in and as all experience. So it would be even more accurate to say, 'Life, like a dome of many-coloured glass, *shines* with the white radiance of eternity.' This is what the Sufis mean when they say, 'There is only God's face.'

If consciousness is the sole reality of all that is, and is itself without form or limitation, how is it possible for a world of form to arise?

In the Gospel of St. Thomas, Jesus says, 'If people ask you, "What sign of your Father is in you?" tell them, "It is a movement and a rest".' The natural condition of consciousness is to simply be at rest within itself, knowing its own being, but it is also natural for consciousness to move or vibrate within itself.

All vibration has a particular amplitude and frequency, and this vibration gives form to the otherwise formless field of pure consciousness. As eternal, infinite consciousness begins to vibrate within itself, thereby assuming a particular form, it ceases to know its own eternal and infinite being and instead appears to itself in finite form. It is similar to the way an empty, 'unlimited' screen appears as a finite image.

This vibrating appearance or finite form of consciousness is experience or mind. As such, we could define mind as the movement or activity of consciousness, rather than an entity in its own right. The only entity there is, if we can call it an entity, is consciousness itself. Mind is consciousness in motion; consciousness is mind at rest. All experience is a movement of mind, and mind – that is, all thinking, imagining, feeling, sensing and perceiving – is a vibration *of* consciousness, appearing *in* consciousness, known *by* consciousness and made *of* consciousness.

Consciousness is to mind or experience as a self-aware screen is to a movie. Just as the only substance present in the limited appearances of the movie is the 'unlimited' screen, so the only substance present in finite experience or mind is infinite consciousness. And just as the screen is never divided into a multiplicity and diversity of objects and characters in a movie, even though it *appears* as such from the perspective of one of the characters, so the only substance present in experience or mind is indivisible, infinite consciousness itself, although it *seems* to comprise a multiplicity and diversity of finite, separate objects and selves from the limited perspective of one of those selves.

The multiplicity and diversity of experience or mind is a modulation of the indivisible, undiversified field of pure consciousness, just as the multiplicity and diversity of the images in a movie are a modulation of the single, indivisible screen. And just as the only substance present in the activity or movement of a character in a movie is the screen, which itself neither acts nor moves, so the only substance present in the activity or movement of mind is consciousness, which likewise neither acts nor moves. Hence, consciousness moves without moving and acts without acting.

*   *   *

If there is nothing in consciousness other than itself, how can it know something other than its own eternal, infinite being, such as an object, person or world?

The infinite can only know the infinite; the finite can only know the finite. One might then wonder, if infinite consciousness and the finite mind never know one another, how is it possible for experience to be known, and by whom, given that infinite consciousness is all that is truly present? In other words, what is the relationship between infinite consciousness and the finite mind? The question is resolved by understanding that infinite consciousness and the finite mind are not two separately existing entities; the latter is a temporary modulation of the former.

In order to know something finite, such as an object, person or world, infinite consciousness must cease knowing its own unlimited being as it is and assume the form of the finite mind. That is, to know an object, person or world that seems to be separate and distinct from itself, infinite consciousness has to divide itself in two and become a finite consciousness or mind, the separate subject of experience, from whose point of view it is able to know a separate object.

Mind is the activity in and as which the inherently unified field of consciousness seems to divide itself in two – a subject that knows and an object, other or world that is known – just as in a dream our own mind seems to divide itself in two, one part becoming the dreamed world and the other the dreamed subject, from whose perspective it subsequently knows the world, although this apparent division of itself never actually compromises its unity.

All experience, however ordinary or extraordinary, is an experience of the finite mind, although the only substance present in the finite mind is infinite consciousness. The finite mind is always changing in appearance, but its essential substance or reality always remains the same, just as the images in a movie are always changing but their reality, the screen, never changes. The original, essential and irreducible nature of mind is infinite consciousness itself. This could also be called original mind or unconditioned mind.

Thus, the mind is not separate from consciousness. In fact, there is no actual thing or entity called 'mind'; there is only consciousness and its modulations. Consciousness can never know anything other than itself, because all there *is* in experience or mind is consciousness itself. In assuming the form of the finite mind, consciousness never actually ceases to be infinite consciousness or to know something other than itself, just as a screen that assumes the appearance of a landscape in a movie never ceases to be the screen.

It is only in the form of mind that consciousness can seem to know something other than its own unlimited, ever-present being, such as an object, other or world. As such, mind is the activity or agency through which and as which consciousness appears to itself as objective experience. To assume the form of mind, consciousness has to narrow the focus of its knowing, thereby seeming to limit itself. Once it has limited itself by assuming the activity of mind, the knowledge of its own eternal, infinite nature seems to be ignored, veiled or forgotten, and this forgetting allows the knowledge of objective experience – that is, manifestation – to come into apparent existence.

Consciousness dreams the world within itself and simultaneously loses itself in its own dream, seeming, as a result, to become a separate subject of experience *within that dream*, from whose perspective the dreamed world may be known. Thus, the subject–object relationship is the medium through which the infinite appears to itself as the finite. The subject and the object are two sides of the same overlooking, ignoring or forgetting of consciousness.

In other words, creation or manifestation requires a sacrifice; it comes at a price. Consciousness must sacrifice itself, must ignore the knowing of its own infinite being, in order to bring manifestation out of itself into apparent existence.

The world of objects is a reflection of the limitations inherent in the perceiving subject. However, if the apparently separate subject of experience explores its own nature by turning the knowing with which it normally knows objective experience upon itself, it is divested of its self-assumed limitations and stands revealed as infinite consciousness. When the apparent subject recognises or 'remembers' itself and thus loses its limitations, its corresponding object is accordingly divested of *its* limitations, and the single, indivisible identity of the subject and object is revealed.

This is what Jesus meant when he said, 'I and my Father are one.' That is, the essential reality of myself – the consciousness that shines in the mind as the knowledge 'I am' or the experience of being aware – and the ultimate reality of the world – existence itself – are a single, indivisible, infinite reality. When the perceiving subject is realised to be an imaginary limitation on the true and only reality of infinite consciousness, the world as an apparent multiplicity and diversity of objects, separate and distinct from the subject, is realised to be equally illusory, and shines as consciousness itself, God's infinite being.

As William Blake said, 'If the doors of perception were cleansed everything would appear to man as it is, Infinite.'* That is, if the mind through which the world is perceived is divested of its limitations, it will cease projecting those limitations onto the object, other or world that it knows or perceives, which will, as a result, stand revealed as infinite. In fact, the object, other or world will cease being experienced as an object, other or world, and will be experienced as the radiance of God's infinite, self-aware being.

*   *   *

To effect the appearance of the finite within the infinite, consciousness must contract or narrow the range of its focus. As such, the full spectrum of experience, from the deepest, so-called unconscious and subconscious layers of mind to the common states of waking, dreaming and sleeping, involves a progressive narrowing, limiting or forgetting of the essential nature of mind, pure consciousness.

* From *The Marriage of Heaven and Hell.*

How does this happen? From the ultimate point of view, it doesn't. Nothing ever happens to consciousness, just as nothing happens to the screen when a movie begins. But let us make a concession to the mind and explore this on its terms. Imagine original mind or pure consciousness, wide open and completely unfocused. There is just its own empty, formless, dimensionless presence, knowing its own eternal, infinite being. In this open and undirected condition, there is no objective experience in the field of its perception, just as a completely unfocused camera will register no image.

However, as consciousness begins to vibrate within itself, its formless being begins to assume the appearance of limitations, and as a result consciousness seems to cease knowing its own infinite being and to know itself instead as limited form. This knowing of form requires a focusing of consciousness, a contraction of its wide-open, unfocused knowing into a limited, directed form of knowing. This limited, directed form of knowing is mind, experience or attention. Thus, it is the focusing or contracting of consciousness that brings form out of potential in infinite being and into finite existence.

To begin with, as consciousness focuses or contracts its knowing in the form of mind, experience or attention, only the most nebulous forms appear, with almost no shape, colour or size. But as consciousness progressively narrows and sharpens its focus, objects become correspondingly clearer. This progression of focus brings into existence a spectrum of experience, beginning with subliminal states of the so-called subconscious and unconscious and passing through the dream state into the waking state, in which experience is at its most clearly defined.

At no stage does a self actually come into existence or move through these states, although, being objective, states can only be viewed or known from the perspective of an *apparent* subject. This subject is not an entity in its own right. It is the *agency* through which consciousness manifests itself as, and simultaneously knows, objective experience.

Consciousness doesn't pass through these states, just as a screen never passes through any of the experiences of a character in a movie. States unfold in consciousness, in the same way that the experiences of the character unfold on the screen. Consciousness is all that is present to know or experience anything, and is, at the same time, the essential nature or content of whatever is known or experienced. Thus, consciousness never enters a state but simply vibrates within itself, unfolding itself within itself. The different states are different frequencies and amplitudes of its own activity.

To say, 'To begin with, as consciousness focuses its knowing in the form of mind' is not strictly true because it suggests that the apparent focusing of consciousness happens at a moment in time. However, time only comes into apparent existence with the movement of mind, so the movement of mind cannot take place or start at a moment in time. Mind doesn't begin or end, because when the movement of mind ceases the appearance of time ceases.

Mind is without beginning or end, but mind itself can never understand that, because the mind imposes its own limitations on everything it knows, thereby making what is, in fact, eternal appear as time. So, as a concession to the mind that is trying to understand its own activity and nature, it is legitimate to say provisionally, 'To begin with, as consciousness focuses its knowing in the form of mind...'

*   *   *

To know form or manifestation, consciousness must focus its knowing or attention in a particular direction. For a specific object – a thought, image, feeling, sensation or perception – to come into the field of experience, consciousness must contract within itself, focusing and thus limiting its knowing in the form of attention. As such, attention brings form into existence out of the formless field of infinite consciousness.

This directing of its attention necessarily involves the exclusion, ignoring or forgetting of everything that is outside its focal field, just as your focusing on these words at present necessarily excludes numerous other experiences, which are, as a result, scattered at the periphery of your field of experience. For instance, the tingling sensation at the tips of your fingers didn't come into existence the moment you read these words. It was there all along, but eclipsed by your interest in these words.

The finite mind, although made only of infinite consciousness, is a narrow segment of the infinite field of pure consciousness, increasing in narrowness as it progressively condenses into the waking state. This does not imply that all that exists are the forms of the dreaming and waking states. The waking-state mind cannot legitimately make that claim, as its own knowledge is, by definition, limited to the appearances of the waking state.

It simply means that as forms become more clearly delineated as they emerge from the infinite field of unfocused consciousness and contract through a spectrum of states into the waking state, more and more of the field of consciousness is excluded and thus relegated to the peripheral and subliminal realms of experience.

In other words, awareness assumes the form of the finite mind by narrowing the focus or sphere of its knowing, progressively excluding the possibilities within its own infinite field as the objects of the waking-state world come increasingly into focus. In order to know itself, a reversal of this process is required, and this is effected through a relaxation of the focusing or directing of its attention. This process can be formalised in the practice of meditation or prayer, but it also happens momentarily when a thought or perception comes to an end – although this is not normally noticed due to its brevity – and as the mind relaxes when falling asleep.

As awareness relaxes the focus of its attention from the object, other or world, its faculty of knowing gradually sinks back into itself and, in doing so, is progressively relieved of the limitations it assumed in order to take the form of the finite mind. At some stage, if consciousness relaxes the focus of its attention all the way back to its essential, wide-open, objectless condition, it remains knowing only itself and knows no object, other or world, just as no image appears through the lens of a completely unfocused camera. There is just infinite consciousness knowing its own inherently peaceful, ever-present and unlimited being. This is experienced in deep sleep, the fulfilment of a desire, the timeless gap between two thoughts or perceptions, a sudden moment of heightened emotion, or a moment of love, beauty or understanding.

Consciousness brings manifestation into existence by freely limiting itself with its own creativity, thereby actualising a segment of its own infinite potential. In other words, consciousness has to fall asleep to its own infinite nature in order to assume the form of mind and, as such, manifest and simultaneously know itself as the world. Thus we could say that when the mind wakes, consciousness sleeps, and when the mind sleeps, consciousness wakes. Of course, consciousness never sleeps; to 'fall asleep to' in this context means to ignore its own infinite reality.

As it says in the Bhagavad Gita, 'That which is day for the many is night for the one, and that which is night for the many is day for the one.' When mind is awake or active, consciousness is asleep to its own nature, assuming the form of the finite mind in order to manifest a part of its infinite potential, that is, to bring the world into apparent existence. When consciousness wakes up to itself and recognises its own infinite being, the mind dissolves or 'sleeps', and as a result, consciousness folds the world up again within itself. There is consciousness knowing its own infinite being, or consciousness veiling itself with its own creativity and appearing as mind.

It is in this context that the world is said to be 'the forgetting of the self' or 'a dream in God's mind'. When God falls asleep, She dreams the world into existence.

*   *   *

How can infinite consciousness fall asleep to itself, or overlook its own infinite nature, when there is nothing in infinite consciousness other than infinite consciousness with which it could be veiled or limited?

Just as in a night dream each of our minds simultaneously creates and identifies itself with a new body, from whose point of view the dreamed world is known, so in the waking state infinite consciousness imagines and identifies itself with our body, from whose perspective it experiences itself as the world. As such, mind is the activity through which and as which consciousness limits and thus localises itself, and the body is an image in the mind of this localisation process. The body is not an *object* made of matter; it is an *activity* made of mind – the activity through and as which consciousness limits and localises itself.

Consciousness veils, ignores or forgets its own infinite nature by freely assuming the limitations of the body. Consciousness falls asleep to its own nature and dreams that it gives birth to the world inside itself, and in doing so seems to become limited to and localised as a body *in that world*, from whose perspective it now seems to know that world. It is for this reason that the mind, whether in the dreaming or the waking state, always seems to know the world from the perspective of a body; it is an illusion that lasts even after the nature of mind has been recognised.

In other words, consciousness enters its own imagination as a separate subject of experience, from whose point of view it knows the world, but in doing so it loses the knowledge of its own eternity and seems, as a result, to become temporary and finite. This is referred to as the Fall in the Christian tradition and ignorance – the ignoring of reality – in the Vedantic tradition.

It is for this reason that in the Vedantic tradition, mind is said to be a mixture of the knowing that belongs to pure consciousness and the limitations that belong to the body. What is often not made clear in this tradition is that this apparent limitation of consciousness for the sake of manifestation is not a mistake that is imposed on consciousness by some malevolent or ignorant external force, but is freely assumed by consciousness itself so as to bring what lies in potential within itself into apparent manifestation. It is an act of love, not an act of ignorance. It is for this reason that William Blake said, 'Eternity is in love with the productions of time.'*

Although consciousness never actually becomes a separate subject of experience, it seems to be such from the perspective of the finite mind. In other words, the ego – the apparently separate subject of experience that seems to be located in and as the body – is only such from its own illusory point of view. It is only as this apparent consciousness-in-the-body entity that consciousness can know something that seems to be other than itself, that is, an object, person or world.

In other words, the world can only be known in duality, in subject–object relationship. Consciousness has to divide itself in two – a subject that knows and an object that is known – in order to manifest creation. It has to sacrifice the unity of its own infinite, indivisible being and seem to become a separate self in the world, which now appears, as a result, to acquire its own independent existence. Thus, the inside self and the outside world are the inevitable duality that constitutes manifestation. They are two sides of the same coin: the apparent veiling of reality.

Being two apparently separate entities, the self and the world seem to have two differing realities. Mind seems to be the essence of the self on the inside and matter is considered, from the perspective of that mind, to be the essence of the world on the outside. Thus mind and matter co-arise as consciousness falls asleep to its own infinite reality and begins to dream the self-in-the-world into existence. As such, neither mind nor matter has its own separately existing, independent reality; both are temporary, finite modulations of the single reality of infinite consciousness.

It is no coincidence that psychiatrists have never found a discrete entity called mind, and physicists have yet to find matter. These substances will never be found, for they do not exist in their own right. Their apparent existence is borrowed from infinite consciousness, God's infinite being, the sole reality of all that is or seems to be.

---

*From *The Marriage of Heaven and Hell*.

CHAPTER 13

THERE ARE NO STATES OF CONSCIOUSNESS

Under the materialist paradigm which presently dominates our world culture, reality is believed to exist independently of mind, and therefore independently of consciousness, the essential nature of mind. From this perspective, matter is the material out of which reality is supposedly made. Mind, which is considered to be derived from matter, is thought to be a limited and distorted view of reality, depending on its state. Consciousness, if it is acknowledged at all, is considered a fleeting and ephemeral quality of mind. Thus, this paradigm starts with the assumption of matter and proceeds to build a model of reality upon it. However, no one ever has experienced, or could experience, matter in the absence of mind, and thus this assumption is founded upon an unverifiable belief.

From the perspective of consciousness, mind is its own activity, never an entity in its own right, and the world is an appearance in and of mind. Thus, from the perspective of consciousness, it is itself the ultimate reality of all experience. From its perspective, no individual mind or world, each possessing its own separate and independent reality, ever actually comes into existence. The world is a condensation in mind, and mind is the activity of consciousness.

Just as a screen never passes through any of the activities or states undergone by a character in a movie, so consciousness never passes through the states of waking, dreaming and sleeping. It is only from the point of view of a finite mind – which believes that it possesses its own independent existence – that a self is believed to move through three states. From the point of view of consciousness, there is no separate, individual self or ego that could transition through any states. Waking, dreaming and deep sleep are temporary modulations of consciousness, which is never itself inherently changed by any of the states it assumes.

Consciousness is to the states of waking, dreaming and deep sleep as a self-aware screen is to emails, images and a blank screen saver, each of which conditions the screen according to its own limitations. Likewise, the three states superimpose their own limitations on consciousness, thereby making consciousness appear in a form that is consistent with their own limitations. And just as the screen is consistently present in the email, image and blank screen saver and never itself undergoes any modification – its modification being one of appearance only – so consciousness remains consistently present throughout the three states of waking, dreaming and deep sleep, and never itself changes.

The three states of waking, dreaming and sleeping are all limited, but consciousness, the underlying reality of the three states, being common to all of them does not share their limits. Even to say that each of the three states is limited is to credit them with too much existence. It is to suggest that each state is a limited object and consciousness its unlimited subject. In fact, each state borrows its apparent existence from the only 'one' that truly is, consciousness itself.

As a concession to the finite mind, it is legitimate to say that the three states are limited, and that consciousness is by contrast not limited, or infinite. However, from the perspective of consciousness, the only real perspective, *it* is the only substance present in any apparent state, so the limitation of any particular state is not a real limitation, only an imaginary one. Thus, the three states are only limited from the limited perspective of one of those states.

*   *   *

The three states of deep sleep, dreaming and waking are progressive modulations of infinite consciousness, each state revealing or expressing a partial and therefore limited view of consciousness's infinite potential.

The true waking state – that is, the state in which consciousness is most 'awake' to its own reality – is consciousness's knowledge of its own indivisible, infinite being, which shines in the finite mind as the experience of being aware, and which thought conceptualises as 'I' or 'I am'. The degree to which such knowledge is forgotten, ignored or obscured by any of the apparent limitations of the three states of deep sleep, dreaming and waking is the degree to which it departs from absolute reality.

The only absolutely true knowledge is consciousness's knowing of its own being, which shines in the mind as the knowledge 'I am' or the feeling of being. All other knowledge requires consciousness to assume the form of a finite mind in order to be known. Consciousness can only arise in the form of the finite mind – or can only precipitate the finite mind within itself – by forgetting, overlooking or ignoring its own infinite reality, and thus the knowledge of anything other than itself is, at best, only relative.

It is for this reason that science will never discover the ultimate nature of the universe on the terms under which scientific enquiry currently operates. The more science explores the universe, the less of a universe it finds! Sooner or later, it will discover that there *is* no universe as it is conceived by thought or experienced by perception. Science consists of a series of perceptions and thoughts. Everything perceived or thought is, by definition, limited. Therefore, science can only know limited knowledge and experience. Although all thoughts and perceptions are made of infinite consciousness, no thought or perception can capture or truly express the nature of consciousness itself, just as the totality of the screen can never appear as an image in a movie.

At some point science will realise that the universe is not a universe, as such. It will recognise that unlimited consciousness is all there is. But who or what could realise that? Only consciousness knows consciousness. Only awareness is aware. Only knowing knows knowing. Therefore, knowledge of the real universe is consciousness's knowledge of itself. 'I am' is the hint in the mind of that knowledge. It is the scent of God's presence in the heart. Hence the Sufis say that whosoever knows their self knows their Lord, that is, whosoever knows what is meant by the word 'I' knows the ultimate reality of the universe.

Until scientists investigate the nature of the subject known as 'I', they will never discover the nature of the universe. Ironically, the very mind that would investigate its own nature must itself dissolve, and it is for this reason that there is so much resistance in most people to exploring, let alone facing, the reality of experience. It is not the nature of the world that is a problem; it is our investment in our own separate identities that makes it so difficult to see the obvious facts of experience.

All knowledge other than the knowledge 'I am' requires a narrowing, limiting or forgetting of infinite consciousness, and therefore can at best be only relatively true. All relative knowledge is true only to the extent to which it partakes of the absolute truth. From this perspective, the waking state, in which reality appears as a multiplicity and diversity of external objects made out of matter, known by an internal self made out of mind, is a kind of ignoring or forgetting of reality. It is for this reason that in the non-dual traditions the normal perspective of the waking-state subject – the apparently separate self around whom most of our lives revolve – is referred to as ignorance.

To be 'ignorant' in this sense is not to be stupid, as the word usually implies in our culture. Rather, it refers to a state of mind in which the reality of experience is ignored, overlooked or forgotten and the corresponding sense of being a separate, finite self, with its accompanying sense of alienation and conflict, seems at its most real. It is in this context that ignorance is, in these traditions, said to be the source of all psychological suffering.

For the waking state to take place – that is, for infinite consciousness to assume the activity of mind and appear to itself in the form of a multiplicity and diversity of objects and selves – infinite consciousness itself must seem to overlook the knowing of its own infinite being and, as a result, seem to become a temporary, finite mind. The body is an image and sensation in the mind of this limiting and localising of consciousness.

The finite mind is infinite consciousness with which the limitations of an apparent body have been mixed. I say 'apparent' because the body is not an object that exists in its own right outside consciousness. Infinite consciousness, pure knowing, modulates itself in the form of sensing and perceiving and appears to itself as a body, in which it limits and localises itself as a finite mind.

It is only from the perspective of that finite mind, apparently living inside the body, that consciousness can know a multiplicity and diversity of objects, called 'the world'. This is what Wordsworth meant when he said, 'Our birth is but a sleep and a forgetting'.* Thus, the true waking up is not to wake up as a body, in a world. It is to wake up as consciousness, knowing its own infinite, indivisible reality alone.

*   *   *

*William Wordsworth, 'Intimations of Immortality from Recollections of Early Childhood', *Poems, in Two Volumes* (1801).

As we fall asleep, the clarity and precision of the waking state begin to dissolve. Attention or mind starts to sink back into its source and is progressively relieved of its defining qualities and thus its limitations. That is why the laws of physics are looser in the dream state than in the waking state, and why the boundaries and divisions between objects and selves are not so clearly defined.

In fact, we don't actually fall asleep or enter a dream state. There is no separate, finite entity or self actually present *in* the dream or waking state. The only entity there is – if we can call it an entity – is consciousness. 'We' are ever-present, unlimited consciousness. In fact, there isn't even a 'we' – a multiplicity of selves – in infinite consciousness: 'we' are considered to be a multiplicity of separate selves only from the illusory point of view of one of those apparent selves, just as the numerous characters that exist in our night dream only seem to have individual existence from the perspective of the apparently separate subject *in* the dream. From the perspective of ever-present, unlimited consciousness itself, it is never divided into a multiplicity and diversity of selves or minds.

Just as, relatively speaking, each individual, limited mind takes the form of numerous thoughts and perceptions but always remains a single, indivisible mind, so consciousness takes the form of numerous finite *minds* but always remains single, indivisible, ever-present and unlimited.

And just as no thought or perception within any apparently individual, finite mind ever acquires a separate status of its own but is always only a modification of the single, indivisible mind in which it arises, so no individual *mind* ever acquires a separate, independent status of its own within infinite consciousness but is always only an undividing modulation of the single, indivisible consciousness in which it appears and out of which it is made.

There are no finite minds! The finite mind is an illusion abstracted by thought. However, all illusions have a reality to them. The reality of the finite mind is infinite consciousness, just as the reality of an image is the screen.

There is only one 'I', the infinite, indivisible 'I' of pure consciousness or God's infinite being, which refracts itself through the activity of its own creativity and appears to itself, in itself, as itself, in the form of an apparent multiplicity and diversity of objects, selves and others. The world is what God's infinite being looks like when viewed from the perspective of an apparently separate subject.

Consciousness never enters a state or becomes anything other than itself. It simply seems to contract and relax, or, more accurately, to focus and defocus like a camera lens. The states of waking, dreaming and sleeping, and any other states that may be experienced, are varying degrees of this focusing and defocusing. When a camera is fully unfocused nothing is seen through it, but as the lens is progressively focused, objects begin to emerge from the unmodulated image, bringing into view what was already present but could not be seen.

Likewise, everything exists eternally in infinite consciousness, in a way that is impossible for the mind to understand. It is the gradual focusing of consciousness – the activity known as mind – that brings the previously unfocused and therefore inaccessible content of consciousness into apparent existence.

From the point of view of the waking state, it is the ultimate reality and all other states prior to it are less real to the degree to which they diverge from it. Thus, the dream state is considered to be an illusion from the point of view of the waking state. Likewise, the subliminal fields of mind that exist prior to the dream state – the personal and collective 'unconscious' – and from which the dream and waking states emerge, are considered, from the limited point of view of the waking state, to be even less real than the dream state.

However, there are no clear lines between any of these states. They are a continuum, appearing in consciousness, known by consciousness and made of consciousness, in which consciousness itself, vibrating within itself, progressively narrows the field of its focus, thereby assuming various forms or states of mind.

As a result of the narrowing of its focus, consciousness seems to become increasingly obscured from itself as the forms of the finite mind become more distinct with the emergence of the waking state, just as a screen seems to be increasingly obscured as an image emerges on it. In other words, there is a price to pay for manifestation. The greater the amplitude and frequency of consciousness's vibration within itself, the more clearly defined, and thus the more apparently separate, are the forms that it assumes.

The spectrum of states that result from this focusing of consciousness ranges from the subtlest archetypal forms that are shared by all minds in the collective unconscious to the apparently discrete forms of the waking-state mind, in which the sense of separation is at its greatest. However, it is only from the point of view of mind that consciousness becomes increasingly obscured by the narrowing of its focus; from its own point of view there is always only ever its own wide-open being.

In the waking state, the separation and otherness of forms, their 'not-consciousness-ness', is at its most persuasive, so persuasive in fact that they seem to be made of an entirely different substance. 'Matter' is the name that thought gives to that substance, defining it as everything outside of and other than consciousness. The fact that no one has ever found, or could ever find, anything outside consciousness has not deterred most people in general and scientists and philosophers in particular from believing matter to be the ultimate reality of the universe, although this is now beginning to change.

The waking state is, as far as we know, consciousness's most contracted or limited form or state, but there is nothing to suggest that there could not be states of mind that are even more contracted than the waking state. We can speculate that if consciousness were to continue to contract beyond the narrow focus of the waking state, the forms in which it appeared would become progressively denser, rendering them completely dark or invisible – a sort of supra-waking state whose contents lie *beyond* it, just as the contents of the dream state, and the personal and collective unconscious, are *prior* to it.

*   *   *

Moving in the other direction, we do not find a clear line between the waking and dream states. The transition from waking to dreaming is a gradual softening or relaxation of the focus of attention. As a result of this softening of focus, the field of possibilities in the dream state is larger than it is in the waking state. Consciousness still needs to localise itself as a body in order to experience the dreamed world, and it is for this reason that we always experience the dreamed world from the perspective of a body in the dream. But in the dream state this self-localisation is looser than it is in the waking state. The world that the dreamed subject experiences is correspondingly less clearly defined and hence there are more possibilities of experience.

In the dream state consciousness has access to a broader segment of its infinite possibilities than it does in the denser, more clearly defined waking state.

As consciousness de-contracts, de-localises or relaxes into itself in the transition from the waking to the dream state, the forms that it assumes become less limited, and thus closer to the 'ultimate form' of consciousness, which has no limits and therefore no definition.

When the focus of consciousness is relaxed in the dream state, consciousness has access to more of the total field of its own activity than it normally does in the waking state. And yet contents from this broader field of mind are continuously making intrusions into the more clearly defined forms of the waking state, often causing disruptions within it. In their mildest form these intrusions may be felt in a positive way as intuition or a deep sense of connection between people, animals and objects. They can also appear in a negative way, as disturbing emotions that seem to arise from the depths of our being, infiltrating and affecting our thoughts, activities, relationships and behaviour in the waking state in a way that is usually beyond our understanding or control.

A similar relaxation or expansion in the field of mind takes place in meditation, enabling previously ignored or suppressed contents from the broader medium of mind to find their way into our experience. It is for this reason that some people's initial experience of meditation is not as pleasant or peaceful as they might have expected! It is only when we move beyond these deeper layers of mind, that is, only when consciousness is sufficiently relaxed, that the inherent peace of our true nature, which lies behind or underneath all these movements of mind, begins to shines through the contents of mind, suffusing it with its perfume of peace and joy.

Just as previously inaccessible regions of the broader medium of mind percolate into our experience in the waking state as it either relaxes naturally as we fall asleep or is induced to by meditation, so in time the peace that lies at the heart of consciousness itself begins to infiltrate the mind, permeating it with its tranquillity and contentment and gradually dissolving out of it the residues of unease, conflict and separation. This is usually felt in the waking state first, but as the peace of our true nature penetrates more and more deeply into the medium of mind, in time the dreaming and deep sleep states are also affected.

A mind that is accustomed to returning regularly to its essence in meditation becomes increasingly transparent to its light and is, as a result, gradually deconditioned from the residues of separation which have dominated it for many years. As a result, agitation, fear, neurosis and the sense of separation and lack give way to corresponding feelings of peace, joy and connectedness to other people, animals and the environment, and in time this change shows its effects in our behaviour, activities and relationships.

These feeling of peace, joy and connectedness are not in fact *new* feelings. They are simply the inherent qualities of our essential nature filtering into objective experience through the previously murky layers of contraction and separation. It is for this reason that we are never motivated to move away from such feelings. They feel like home. They feel like our birthright. They are our birthright! The recognition of our true nature – its recognition of itself – will always decrease our suffering, although the extent to which and the speed at which this happens will be dictated in most cases by the density of our previous conditioning.

The body, being a denser vibration of mind than are our thoughts and feelings, usually takes longer to be permeated by the peace of our true nature, but in time even the body begins to be colonised by the light of consciousness, leading to a profound relaxation and an increase in sensitivity and openness. When a well-known Zen master was asked by one of his students, whilst lying on his deathbed, 'How is it for you now, Master?' he replied, 'Everything is fine, but my body is having a hard time keeping up!' Although his mind was clear, in his wisdom and humility the Zen master recognised that there were still some residual pockets of experience that were yet to be colonised by the light of consciousness.

The deeper layers of the separate self are laid down as a network of feelings and sensations that permeate the body, imbuing it with a feeling of heaviness, opacity, insensitivity, dullness, depression and, above all, separation from others and the environment. In most cases these feelings of heaviness and separation survive the recognition of our true nature, and a further process is required to expose and dissolve them. Such feelings are immune to rational analysis, and therefore the written word is not the ideal medium for their exploration.*

In meditation the mind becomes sensitive and open; at least, that is the point of view of the individual. It would be more accurate to say that in meditation the activity of consciousness – mind – relaxes, the forms that it assumes become correspondingly less clearly defined, and thus the boundaries between these forms become less obvious and in time dissolve

---

*I refer anyone who would like to make a deeper exploration of these residues of separation that remain hidden in the body to my collection of meditations, *Transparent Body, Luminous World – The Tantric Yoga of Sensation and Perception*, published by Sahaja Publications.

completely. It is for this reason that long-term meditators often develop a degree of sensitivity towards objects, animals and others that expresses itself in loving and compassionate ways.

As meditation deepens, the inherent transparency and sensitivity of consciousness is no longer dulled or obscured by its own activity. People and objects become transparent to the light of consciousness, no longer veiling consciousness but shining with it, announcing its presence and delivering something of its beauty, which is beyond the mind's capacity to know or grasp.

CHAPTER 14

WORDSWORTH AND THE LONGING FOR GOD

Although the non-dual understanding was more freely expressed in the East than it was in the West, due to its suppression by the church, there are many instances of it in Western literature. One such example is William Wordsworth's poem, 'Intimations of Immortality from Recollections of Early Childhood',* in particular the fifth stanza:

> Our birth is but a sleep and a forgetting:
> The Soul that rises with us, our life's Star,
> Hath had elsewhere its setting,
> And cometh from afar:
> Not in entire forgetfulness,
> And not in utter nakedness,
> But trailing clouds of glory do we come
> From God, who is our home:
> Heaven lies about us in our infancy!
> Shades of the prison-house begin to close
> Upon the growing Boy,
> But he beholds the light, and whence it flows,
> He sees it in his joy;
> The Youth, who daily farther from the east
> Must travel, still is Nature's priest,
> And by the vision splendid
> Is on his way attended;
> At length the Man perceives it die away,
> And fade into the light of common day.

*From *Poems, in Two Volumes.*

Wordsworth's use of the image of the birth of an infant and its subsequent growth into adulthood can be taken as a poetic analogy of the birth of the finite mind from unlimited consciousness. This cycle from infancy to adulthood could also be applied to the twenty-four-hour cycle of the emergence and dissolution of the waking state. As such, 'Our birth' could be understood as the emergence of the waking state from the field of infinite consciousness, which Wordsworth describes as 'a sleep and a forgetting'.

Consciousness is by nature self-aware. Its nature is simply to know its own eternal, infinite, inherently peaceful and unconditionally fulfilled being. However, the infinite can only know the infinite; the finite can only be known by the finite. Therefore, in order to know a limited form or manifestation, consciousness must seem to cease being infinite and assume the form of the finite mind, the limited subject or knower from whose perspective objective experience may be known. Consciousness assumes the form of the finite mind by limiting itself to and locating itself in the body. As such, the body is the agency through which infinite consciousness becomes a finite mind or separate self. The name that the separate self gives to itself is 'I'.

It is no coincidence that although all separate selves are known by others by a variety of names, all separate selves know *themselves* as 'I', a subliminal recognition of the fact that at the heart of all finite minds shines the same infinite, indivisible consciousness, of which all apparently separate selves are but partial reflections. It is for this reason that the first true statement that all apparently separate selves can make about themselves is simply 'I am', a statement that confesses the knowing of being that shines in all minds at all times, irrespective of circumstances, situations or states of mind. And it is for this reason that the knowledge 'I' is considered a portal through which the separate self passes on its return journey back to its home in pure consciousness, and the same portal through which infinite consciousness passes in the other direction as it assumes the form of the finite mind.

This apparently limited self or ego is the 'Soul' to which Wordsworth refers. It is the beacon of consciousness around whom the finite mind revolves, that shines – 'our life's Star' – as 'I' in the midst of all experience. However, this 'I' does not belong to, nor does it share the limits or destiny of, the finite mind. The finite mind borrows the knowing with which it knows its experience from infinite consciousness, the only consciousness there is.

The soul or separate self 'hath had elsewhere its setting and cometh from afar'. It 'comes from' infinite consciousness. It is a contraction within eternal, infinite consciousness or, in religious language, God's infinite being. The soul or separate self is a crystallisation of the infinite field of pure consciousness, a dream in God's mind, from which it derives its being and its sense of 'I'-ness.

However, in the early stages of this emergence, fresh out of infinite consciousness, the separate self is still saturated with its fragrance, that is, saturated with God's presence. Hence Wordsworth says, 'not in entire forgetfulness, and not utter nakedness, but trailing clouds of glory do we come from God, who is our home'. As the finite self starts to crystallise out of God's infinite being, it retains the memory of its own eternity embedded within it.

That is why, as the thin veil of nothingness that shrouds consciousness's knowing of its infinite being in deep sleep begins to diversify and multiply itself as the dream and waking states, these states are still transparent to the light of consciousness and thus saturated with its peace. It is for this reason that the early morning, before the forms of the waking state have fully crystallised, and the evening, as their apparent solidity is beginning to dissolve into the fluidity of the dream state, are considered auspicious times for meditation, when the natural cycle of emergence and dissolution are cooperating with the mind's longing to sink into its source.

In the waking state, all minds, feeling the immanence of the freedom and peace from which they have emerged, long to be divested of their limitation or separateness and return to their 'home' of infinite consciousness, God's being. This longing to be divested of its limitations is the desire that the apparently separate self feels for happiness, intimacy and love.

*   *   *

Wordsworth continues to describe the emergence of the finite mind as a further contraction or restriction of infinite consciousness, a deeper forgetting or ignoring of its own eternal nature: 'Shades of the prison house begin to close upon the growing Boy'. And yet, at this stage in the emergence of the waking-state mind, consciousness's knowing of its own eternal being still filters into experience as repeated moments of happiness. Experience is still transparent to and pervaded by God's infinite being, from which it flows: 'He beholds the light, and whence it flows. He sees it in his joy'.

The boy's feeling of happiness is the experience of the free and inherently fulfilled light of infinite consciousness that is modulating itself in and as his own mind. However, it is not in fact the *boy* who experiences happiness, God's infinite being; a person cannot experience God's being. Only God's infinite being can know God's infinite being. Only infinite consciousness can know infinite consciousness, for there is no other self or entity present, either to know or not know it. The person is simply an imaginary limitation of the infinite and only being there is.

As Balyani said, 'No one sees Him except Himself, no one reaches Him except Himself and no one knows Him except Himself. He knows Himself through Himself and He sees Himself by means of Himself. No one but He sees Him.'

To know itself, consciousness does not need to rise in or assume the form of mind. It knows itself by itself, in itself, as itself. The boy's mind, which knows itself as 'I', is a localisation or contraction of consciousness, the true and only 'I', which consciousness itself freely assumes in order to actualise or realise a segment of its own infinite potential in the form of the boy's experience of the world. The boy is, as such, a process or an activity made out of mind, not an object made out of matter, through which consciousness filters itself in order to experience a world. The boy is the agency through which God knows itself as the world.

The emergence of the deep sleep, dreaming and waking states out of infinite consciousness could be seen as a progressive veiling of consciousness. In deep sleep there is a thin veil of nothingness or blankness over the inherent peace of our true nature. This veil is not sufficiently thick to eclipse the inherent peace of our true nature, and that is why deep sleep is experienced as peaceful.

In the dream state this veil begins to diversify itself and, as a result, appears as a multiplicity and diversity of loosely arranged objects and selves, which only become fully concretised as apparently independent entities in the waking state, just as the screen of a laptop could be said to first thinly veil itself in the form of a blank screen saver, which subsequently diversifies itself into emails and images.

The waking state, in which objects and selves are at their most distinct and separate, could be said to be the farthest from consciousness as it progressively contracts within itself and assumes the forms of experience. Likewise, the dissolution of objectivity as we progress from the waking state to the dream state, and from the dream state to deep sleep, could be seen as a progressive thinning out of consciousness's veiling of itself.

However, this is said only as a concession to the separate self that seems to come into apparent existence as the dreaming and waking states emerge. From the perspective of consciousness – and, of course, consciousness's perspective is the only one that is real – no state veils itself in the slightest degree, in the same way that from the point of view of a self-aware screen no image, however dark, agitated or diverse, obscures it in any way.

In other words, the veiling of consciousness is only real from the limited and ultimately illusory perspective of a separate self that seems to come into existence with the emergence of these states. Thus, ignorance is only for the apparently separate self, never for consciousness, and it is for this reason that in the Vedantic tradition ignorance is said to be unreal. It is, as such, referred to not as ignorance but the illusion of ignorance.

*   *   *

The word *maya* is used in the non-dual traditions to describe consciousness's ability to assume a form with which it seems to limit itself. It is the power that a screen possesses to appear as a landscape and, as such, seem to veil itself with its own creativity. From this perspective *maya* is often translated as 'illusion', that is, the ability of infinite consciousness, the self-aware screen, to appear as something other than itself, which it now knows from the perspective of a separate subject within its own dream. However, the illusion is only such from the limited and ultimately imaginary perspective of the separate subject of experience that seems to come into existence as a result of consciousness's veiling power.

Maya, as illusion, is the activity of mind through which infinite consciousness brings manifestation out of its own being into apparent existence. It is its own cause. However, from the point of view of consciousness, its ability to assume innumerable names and forms does not create the *illusion* of a world, but is rather seen and experienced as an ever-changing outpouring of itself within itself in order to realise, manifest and enjoy the endless flow of its own infinite potential in form. Thus, the deeper meaning of the word *maya* is 'creativity', the process by which consciousness manifests itself as an ever-changing flow of experience without ever ceasing to be and know itself alone.

In other words, the veiling of consciousness is only such from the perspective of the separate subject of experience. From the perspective of a separate self, *maya* is an illusion; from the perspective of consciousness, it is an expression of its own inherent freedom and creativity, with which its never-changing reality appears in the form of ever-changing experience. As Balyani said, 'His veil is His oneness since nothing veils Him other than Him. His own being veils Him. His being is concealed by His oneness without any condition.'

Thus, when the apparently separate self is divested of its self-assumed limitations and stands revealed as the true and only self of infinite awareness, *maya* ceases to be a veiling power and is experienced as a revealing power, and in correspondence with this change, objective experience, which once seemed to veil consciousness, now shines within it.

Consciousness knows itself in and as the totality of experience. Even our darkest moods shine with the light of its knowing. This inability of consciousness to be, know or become anything other than itself is the experience of love, which admits no separation, objectivity or otherness. Thus, from the perspective of consciousness, creation is a manifestation of love.

*   *   *

In order to manifest itself as a world, consciousness contracts within itself, sacrificing the knowing of its own infinite being for the sake of love, and in doing so seems to cut itself off from its own innate peace and happiness. For this reason there is a wound in the heart of all apparently separate selves, which most seek to alleviate by losing themselves in objective experience.

In order for consciousness to reclaim its innate happiness, a reversal of this process takes place, and it is for this reason that in the Tantric tradition it is said that the path by which we fall is the path by which we rise. The experience of happiness is the relaxation of this process, the unwinding of the self-contraction. This reversal is the de-localisation of awareness in which the mind is gradually, in most cases, divested of its temporary limitations and returns to its original, unconditioned, irreducible essence of pure consciousness, which shines as happiness itself.

In fact, the mind doesn't return to its original condition; it never left there. There is nowhere for the mind to reside other than its own nature of unlimited consciousness. It would be more accurate to say that consciousness ceases to veil itself with its own creativity, and is revealed to itself as it is. It recognises itself. The knowing of its own ever-present, unlimited being *is* the experience of happiness itself, and it is for this reason that all apparently separate selves seek happiness above all else.

The dissolution of the mind's limitations, which is itself the experience of peace or happiness, happens naturally and gradually as the waking-state mind gives way to the dream state, and as the dream state dissolves in deep sleep. In deep sleep only a thin veil of nothingness obscures awareness's knowing of its own unlimited being; it is sufficiently transparent to afford the mind a measure of peace or happiness.

This dissolution also takes place momentarily on the fulfilment of a desire, when the mind's activity of seeking comes briefly to an end and, as a result, the mind plunges into its source and briefly tastes the unconditional peace and inherent fulfilment of its true nature. After this experience of peace or happiness, the mind, on rising again within the ocean of consciousness, usually attributes the fulfilment that it experienced to the object, substance, activity or relationship that preceded it, and therefore seeks the same experience again.

Although these brief moments give the mind samples of the lasting peace and happiness it desires, they never fully satisfy it. At some point it begins to dawn on the mind that it is seeking peace and happiness in the wrong place. This intuition may occur spontaneously as a result of repeatedly failing to secure happiness in objective experience, or as a result of a moment of despair or hopelessness when the mind, having exhausted the possibilities of finding fulfilment in objective experience, finds itself at a loss and, with no known direction in which to turn, stands open, silent and available. In this availability the mind is receptive to the silent attraction of its innermost being, drawing it backwards, inwards or selfwards, a call that is always present but usually obscured by the clamour of its own seeking.

The unwinding of the mind may also be effected in more extreme moments of great fear, sorrow or loss, when the coherence of the mind is temporarily disturbed and it is 'thrown back' into its original condition, a fact that the Tantric traditions have developed into a series of formal practices in which the mind surfs intense emotion back to the shore of awareness. It can also be brought about in moments of heightened pleasure, such as sexual intimacy, when the mind is expanded beyond its customary confines by the intensity of the experience and, as a result, tastes the nectar of its own immortality.

In fact, from this perspective the experience of pleasure, normally the enemy of spiritual realisation in the religious traditions, is considered a taste of pure consciousness. In the moment of aesthetic pleasure, the wandering mind is brought to bear so intimately on the object of perception as to merge with it. In this merging the mind briefly loses its limitations, and its essence of pure consciousness shines. That is the experience of beauty. It is the experience that the artist seeks to evoke, and to which Paul Cézanne referred when he said that he wanted his art to give people the taste of nature's eternity.

It is also the experience with which the lover seeks to unite. In her poem *Woman to Lover,** Kathleen Raine describes the dissolution of the separate self which is the essence of all intimacy:

> I am fire
> Stilled to water
> A wave
> Lifting from the abyss
> In my veins
> The moon-drawn tide rises
> Into a tree of flowers
> Scattered in sea-foam
> I am air
> Caught in a net
> The prophetic bird
> That sings in a reflected sky
> I am a dream before nothingness
> I am a crown of stars
> I am the way to die

This dissolution can also be solicited, invoked or fostered in meditation or prayer; likewise through a conversation or a passage in a book, through words that are informed by and infused with its silence. Or it may be precipitated by a question such as, 'Are you aware?', 'What is the nature of the one you call "I"?' or 'What is the nature of the knowing with which you know your experience?' Likewise, the mind may be drawn spontaneously into its source of unlimited consciousness simply by the silent presence of a friend in whom the recognition of their true nature has taken place, without the need for conversation.

*Kathleen Raine, *Collected Poems*, Counterpoint Press (2001).

The mind may also be divested of its limitations by the silent or explosive power of a work of art whose creation was informed, knowingly or not, by the mind's recognition or intuition of its essential nature. Such a work bears the signature of its origin, which it dispenses freely for all who have eyes to see or ears to hear. Suffice it to say that the mind receives numerous hints as to the source of the peace and happiness for which it longs, although these hints, in the absence of correct guidance, usually pass unnoticed.

* * *

Wordsworth continues his poetic description of the birth of the finite mind from unlimited consciousness: 'the Youth, who daily farther from the east must travel, still is Nature's priest'. The east is the source of light, unlimited consciousness. We are in the dream state now. Consciousness is narrowing the focus of its attention and, as a result, form is becoming more clearly delineated and diversified at the expense of the infinite possibilities latent in pure consciousness itself: 'At length the Man perceives it die away, and fade into the light of common day.'

The 'light of common day' is the full emergence of the waking state, where consciousness's apparent overlooking, ignoring or forgetting of its own being, and the subsequent division of itself into an inside self, made out of mind, that knows, and an outside world, made out of matter, that is known, is at its most persuasive. In the waking state, consciousness's knowing of its own eternal being fades as it is eclipsed by the glare of the waking state. Consciousness falls asleep to its own reality for the sake of its creation.

The waking state is like 'a sleep or a forgetting' in which the multiplicity and diversity of objective experience seems to veil the consciousness that is its sole reality. Although consciousness seems to lose itself in its own creativity, this forgetting is not a mistake. It is the agency of manifestation, the activity of creation, just as an actress has to lose herself in the character in order to play the part fully. Once consciousness has lost itself in its own creativity and assumed the form of the 'Soul' or separate self, it then has to embark on a great journey to recover its lost freedom and happiness. Although that journey is initiated by the separate self, it ultimately requires the loss of its separate identity. It requires the death of the very self that seeks freedom and happiness.

Manifestation is simultaneously consciousness's sacrifice of itself and its celebration of itself. The same process of self-sacrifice and celebration is performed in the microcosm of an artist's studio, where the artist has to surrender herself entirely in order to bring forth what is within her. Every artist will recognise this process in the words of the Sufi mystic and poet Hafiz: 'I was a hidden treasure and I longed to be known.' This longing lives in the hearts of all people and initiates the process of creation that has taken place in caves, studios, theatres and concert halls since the dawn of humanity.

CHAPTER 15

THE SHARED MEDIUM OF MIND

The states of waking, dreaming and deep sleep are only such from the perspective of an apparent entity in one of those states. From the perspective
of consciousness itself, which is the only true perspective, it is always in the
same condition, accessing a spectrum of its infinite possibilities in the form
of various states but never entering any state itself, just as a screen is always
in the same condition irrespective of the programmes that are playing on it.

Because the laws of physics are looser in the dream state than they are in
the waking state, corresponding to the degree of relative relaxation of consciousness, events that would be considered magical from the point of view
of the waking state are quite possible in our dreams. For instance, in a
dream consciousness can localise itself in and as a body on the streets of
New York one moment and be in Paris, in and as another body, the next.

Likewise, due to the fact that all finite minds are precipitated within the
same field of infinite consciousness, and that each finite mind is without
a clear boundary, communication between minds, as well as between
states in any one mind, is equally possible. Telepathy, synchronicity and
intuition are all examples of the normal boundaries of the waking state
becoming relaxed and the boundaries between finite minds becoming
correspondingly looser.

Such experiences are embarrassing inconveniences under the materialist
paradigm, but they occur too frequently to be dismissed. They are only
strange from the materialist point of view, which considers the evidence
of the senses in the waking state to be the ultimate arbiter of reality.
Indeed, Einstein is supposed to have said that common sense is a collection of prejudices acquired by the age of eighteen.

From the point of view of the consciousness-only paradigm, in which all experience is understood as a contraction and relaxation of consciousness in the form of the finite mind, there is nothing extraordinary about such occurrences. Indeed, they are to be expected. In this model it is no stranger for there to be communication between minds at a distance than it is for objects to be connected in physical space. The physical space that apparently connects objects is a pale reflection at the level of the finite mind of the true medium in which all experience appears and through which it is connected: consciousness alone.

Finite minds are localisations within a single field of infinite consciousness, with no distinct boundaries between them, or between states within any one mind. When minds are informed by the same frequency of vibration, they experience a shared world. When a mind does not share a particular frequency with any other mind, the resulting experience is private.

At the deepest level all minds are connected because they are all precipitated within the same field of infinite consciousness, and the varying degrees of connectedness that we feel with one another or with animals, objects and nature are the degrees to which our minds are transparent to this shared medium. Love is the word we use when we *feel* this shared medium with other people and animals. The same experience is referred to as beauty in relation to objects.

*   *   *

The body is an appearance of mind, which is itself a modulation, colouring or conditioning of infinite consciousness, so the death of the body entails the dissolution or unravelling of a particular configuration of mind. However, there is nothing to suggest that the underlying forms and energies of mind, which previously condensed to appear as the body, may not remain in a looser configuration when the body disappears.

The disintegration of a whirlpool leaves a residue of ripples in the river in which it was localised long after the whirlpool has lost its specific form, and this residue may form the basis on which a new whirlpool coalesces downstream.* In the same way, the forms or energies of mind that coalesced to form the appearance of a body in one lifetime may remain

*This image is borrowed from Bernardo Kastrup.

140

present in consciousness after the body has disappeared, and there is nothing to suggest that these residues may not coalesce 'downstream' and appear in the form of a new body, and thus in another life.

This 'new body' – which is a new appearance of mind – will bear traces of the forms that were present in the previous body for the simple reason that the energies that were released when the latter disintegrated will go into the formation of the next as it takes shape. This phenomenon gives rise to a theory of reincarnation that is consistent with the consciousness-only model, without making a concession to the idea of a reincarnating entity or self.

The same model also explains why elements of the waking state remain present in an altered form in the dream state. It is our common experience that the emotional and psychological residues of the waking state remain in consciousness when the waking-state self and its world disappear in sleep, and form the basis of the subsequent dream. These residues coalesce in the dream state to form not only the character of the dreamed subject but also its environment, that is, not only its internal but also its external experience.

In this way, what was present as thoughts and feelings *inside* us in the waking state may become our *outside* environment in the dream state. For instance, one with a fearful disposition in the waking state may find herself being chased by a tiger in the dream state. The tiger is an outward manifestation in the dream state of the fear that was within her in the waking state.

The waking state is only a narrow segment of the activity of mind, through and as which consciousness brings a segment of its infinite potential into apparent existence. As consciousness relaxes, its field of focus widens and, as a result, it has access to a broader segment of its own infinite field of possibilities. Whether we call this broader field of mind the collective or personal unconscious, altered states, drug-induced states, out-of-body experiences or near-death experiences, there is obviously much more to mind than is ordinarily accessible during the waking and dream states.

However, none of these states of mind, however ordinary or extraordinary, ever take place outside consciousness. Nor is it quite right to say that they take place *within* consciousness, as if a state of mind were one thing and consciousness another. All states of mind are *modulations* of the ever-present, underlying reality or original nature of mind, pure consciousness itself.

Everything there is exists in and as a modulation of consciousness, although not everything there is always appears in the narrow focus of the waking-state mind. Even now, if I were to draw your attention to the tingling sensation at the soles of your feet, you would suddenly seem to become aware of that sensation. The experience was, in fact, already in consciousness but, due to the exclusive focus of your attention on these words, seemed to be obscured by them.

The tingling sensation was in consciousness all along, but not in the narrow field of the waking-state mind until it was pointed out, at which point attention shifted to the sensation, thereby illuminating it. The experience was always present but previously unknown, just as a torch in a dark room illuminates objects that are already present but were previously unseen. In this way, the focusing of attention brings a segment of the total field of consciousness into view at any one time by excluding all other possibilities. Consciousness rises in the form of attention precisely for the purpose of collapsing its field of infinite possibilities into a single actuality.

* * *

This mechanism accounts for the phenomenon of memory under the consciousness-only model. Consider the situation when somebody's name is on the tip of your tongue. 'On the tip of my tongue' means 'I know it but I don't know it in this moment'. In that moment, where is the friend's name? If we know it, it must be in consciousness, the only 'place' where any knowledge or experience can reside. The reason we feel that we don't know it is that at that moment the waking-state mind has no access to that information. It is in consciousness but outside the limited compass of the waking-state mind.

Moreover, the harder we try to think of our friend's name, that is, the more we focus on it, the more it eludes us, until we intuit that to remember the name we must relax the focus of attention rather than concentrate it. We feel that the name is there but we don't have access to it, and our intuition is correct. The name *is* there, just outside that small part of the field of consciousness that has been brought into focus in the form of the waking-state mind. As the waking-state mind relaxes, so its focal field widens, allowing a larger segment of the field of consciousness to come within its compass. And sure enough, as the mind relaxes or defocuses, contents that were previously outside its sphere are now experienced within it... 'Sophie!'

The name Sophie doesn't appear from somewhere outside consciousness. It is always in consciousness, the only place it is possible for anything to be. Nor does it move from one segment of consciousness – the unconscious – to another – the waking-state mind. The name always stays in the same place, the only place it is possible to be, the placeless place of consciousness. As the mind relaxes its self-contraction it encompasses more of the field of infinite consciousness, in the way that an inflating balloon has increasing access to the space into which it is expanding. As a result, what previously seemed to be *outside* mind is now experienced *inside* it.

Mind conceptualises this experience as memory, and conceives the apparent distance between the name that was unknown and the name that is now known as time. However, both the previously unknown name and the name that is subsequently known appear in the *same* dimensionless consciousness, for which all experience is *now*. Thus, time is not validated by memory; memory, in the form of thought, creates the illusion of time. It is the activity of consciousness, in the form of the apparent expansion and contraction of mind, that makes what is eternally present now appear as a succession of events in time.

* * *

A physical object is an experience of mind – seeing, hearing, touching, tasting and smelling – not an object made out of matter. In order to experience an object, infinite consciousness collapses into subject–object relationship, experiencing the apparent object from the limited perspective of a subject. An object is, as such, a crystallisation of the broader medium of mind in which it appears, and it is for this reason that the apparently solid object retains and expresses the vibrational qualities of the field out of which it emerges. It is like an imprint from the larger field of mind onto the waking state.

This is perhaps most obvious when we listen to music, in which the form of the music – sound – is a direct transmission of the field of mind out of which it emerges. But the same is true of objects that appear in a more solid, concrete form, such as physical works of art. A work of art brings into the focus of the waking state, from the broader, shared medium of mind, knowledge that is normally inaccessible to it, and makes it available to humanity.

Carl Jung called this shared field of mind the 'collective unconscious', which is a somewhat misleading term in that it implies that the contents of this field lie outside consciousness, which is not the case. It is, rather, the collective field of consciousness which lies, for the most part, outside the compass of the waking-state mind and makes itself known to the individual mind through dreams, images, intuitions, and so on.

The content of this broader, shared medium of mind belongs to everyone, that is, to the consciousness that informs all finite minds, but in order to manifest as an object of knowledge or experience it has to be filtered through the prism of a *particular* mind. Shakespeare describes this process: 'And as imagination bodies forth the forms of things unknown, the poet's pen turns them to shapes and gives to airy nothing a local habitation and a name.'* Each finite mind brings a segment of infinite consciousness, 'airy nothing', into actualisation, thereby giving its otherwise unknowable, formless reality a local name and form.

The function of an artist is to bring into the field of the waking-state mind knowledge that comes from the broader medium of mind in which it is precipitated but to which, under normal circumstances, it has no immediate access. The result is a work of art that inspires humanity to a life of love, beauty and understanding. As such, the artist is a function, not a person. It is the function within humanity that serves to restore the balance where separation, despair, conflict and hostility have eclipsed the light of love and understanding that lives in each of our hearts. Art is remembrance.

I first recognised this when, as a ceramic artist, I would visit museums around the world and explore their collections of early pottery. Long before I was able to rationalise experience as I am doing now, I would frequently feel an uncanny familiarity with a particular bowl or jar, a sort of visceral intimacy that expressed itself in simplistic terms such as, 'I know the person who made that bowl', 'I made that jar myself' or 'These are my friends'. I was experiencing what the French poet René Char called 'the friendship of created things'. I was recognising the broader field of mind

*William Shakespeare, *A Midsummer Night's Dream* (c. 1597).

that I shared with the bowl or jar, of which my body and their forms were, as it were, partial representations. Indeed, it was something about the visual image of the bowl or jar itself – my only experience of which was a perception in and of the mind – which had the power to draw my mind away from the objective aspects of experience, through subtle layers within its own field, at least some way 'back' to its formless source and essence.

Seen in this way, such an object becomes, as it were, transparent, delivering to one's intimate experience the broader field of mind of which it is a temporary, local expression, and at some point dissolving the finite mind in the source of pure consciousness from which it emanates. This apparent merging of the field of the perceiver with the field of the perceived is the experience known as beauty. In fact, it is not a merging of two fields but rather the dissolution of apparent distinctions within the essentially indivisible field of their shared continuum. Such is the function and power of art, the power that some objects have to draw attention from the finite to the infinite. In this way, the experience of beauty is a communication of truth, an intervention of reality into the world of appearances.

The same experience can be felt between people and with animals, only in this case it is referred to as love rather than beauty. It is only in the narrow segment of consciousness known as the waking state that minds, and therefore people, seem to be separate from one another. However, if minds were truly separate from one another, the experience of love, or even friendship, would not be possible. Love is the experience of our shared reality. It is no coincidence that people value love above all else.

A person is not an object made of matter; a person is an activity of mind, a field of experiencing. A person, in the conventional sense of the word, is how that activity appears from the perspective of another finite mind, how the object looks from the perspective of a subject. From the inside, none of us has the experience of being a solid, well-defined object. We know ourself rather as a field of knowing or experiencing, vibrating within itself in the form of all thinking, feeling, sensing and perceiving, but never actually condensing into a solid, permanent object or self.

This field of knowing or experiencing is always tending towards the dissolution of its limitations, the dissolution of any force that would tend to capture, suppress or limit it. The longing for freedom, love, peace and happiness that lies in the hearts of all apparently separate selves is only the longing for this dissolution. Friendship is both a catalyst for and an expression of this dissolution.

The term 'satsang' – from the Sanskrit *sat*, meaning 'being', and *sangha*, 'community' – has been downgraded by the contemporary Neo-Advaita movement to indicate a talk in which a speaker informs students. Originally the term had a deeper and subtler meaning, suggesting that the sharing of being is the vehicle of this dissolution. In the New Testament, the same understanding is expressed by St. Matthew, 'For when two or three are gathered in My name, there am I in the midst of them.' In this gathering, the shared essence of each of our minds is magnified and shines as the experience of love.

At the level of the waking state, our bodies and minds appear to be separate, as do the characters in a movie. But just as characters in a movie are modulations of the same indivisible screen, so our waking-state minds and the bodies that appear in them are energetic emanations from the same indivisible field of mind whose nature is infinite consciousness.

* * *

A question that is commonly asked after encountering this approach is, 'If everything appears in the same consciousness, and I am essentially that consciousness, why am I not aware of everybody else's thoughts and feelings?'

Although each finite mind experiences only its own contents, it is at the same time precipitated within the shared medium of infinite consciousness, of which it is a cross-section or partial view. Each of our finite minds brings a segment of infinite consciousness's potential into actuality. As such, each of our minds could be considered a sphere or field that emerges in a shared, self-aware space, focusing the potential that exists unmanifest within it.

When two spheres overlap they share the overlapping part of their content; that part of each sphere that does not overlap with the other is particular to that sphere alone and is experienced as its own private content. Thus it is possible for all minds to be precipitated within the same field of consciousness and for some of their content to be shared – what we call the world – and some of it to be private – that is, thoughts and feelings.

Just as each of our thoughts, sensations and perceptions is the product of a single mind, there is nothing to suggest that each of our minds is not itself the product of a single consciousness. In other words, just as there is a consistency to all our own thoughts, sensations and perceptions precisely because they are all a product of the *same* finite mind, it is not unreasonable to expect that there will be a consistency across finite minds – the experience of a shared world – simply because they are all the product of the *same* infinite consciousness. That is, it is the fact that consciousness is shared between minds that accounts for our experience of a shared world. The world is shared because consciousness is shared!

Moreover, although there is an obvious correspondence between some of the contents of our own minds – for instance, there is clearly a connection between the thought 'What is two plus two?' and the answer 'Four' – there is no obvious correspondence between other such contents, such as the thought that we are currently having and the memory of last year's summer holiday. Likewise, it is to be expected that large areas of our experience may overlap considerably with other minds with whom we are in close contact, for instance, family, friends and neighbours; to a lesser extent with those with whom we are not in contact, for instance, someone living on the other side of the world; and not at all with those with whom we have no contact whatsoever, such as a frog or a snail.

However, just as the apparent disconnection between the elements that appear in a single mind never compromise the integrity of that mind, so the fact that some minds do not share their content – for instance, two people living on opposite sides of the earth – or the fact that two people in close proximity cannot know one another's thoughts, doesn't prove or imply that the ultimate reality of each of these minds is not the same shared medium of consciousness.

All there is to the finite mind is consciousness, but there is much more to consciousness than the finite mind. This is often misunderstood in contemporary expressions of non-duality, which mistake the non-dual understanding for solipsism. Solipsism is the belief that only the content of 'my' finite mind exists; it is a form of insanity. In the non-dual perspective it is understood that only *consciousness* is, and that everything appears in, is known by and is ultimately made of that very consciousness.

Thus the consciousness-only model does not preclude an *apparent* multiplicity and diversity of minds, but recognises that this apparent multiplicity and diversity doesn't actually multiply or diversify consciousness itself. In just the same way, each of our finite minds is capable of a multiplicity and diversity of thoughts and perceptions without compromising its single, indivisible status. In other words, an apparent multiplicity and diversity of minds is only such from the limited and ultimately illusory perspective of *one* of those apparent minds. Just as the thought or perception that each of our minds is now experiencing is only a fraction of the total possibilities that exist within that single, indivisible mind, so each finite mind is itself just a fraction of the total possibilities that exist within single, indivisible consciousness.

*   *   *

Instead of imagining objects and people made out of matter that supposedly exist outside the field of consciousness, know and feel that all that is experienced is a vibrating field of mind, which is itself a modulation of pure knowing or consciousness itself. This vibrating field of mind appears in the form of objects and people on the outside, and thoughts, feelings and images on the inside, including the appearance of our own body.

It is reasonable to infer that other people's experience appears in the same way as our own, that is, as a single field of vibrating mind taking the forms of thinking, feeling, sensing, seeing, hearing, touching, tasting and smelling. When I say 'other people's experience' I do not mean to suggest that people *have* experience. People *are* experience! Only consciousness 'has' experience. Only consciousness *is* experience. These vibrating fields of experiencing are all precipitated within the single, dimensionless field of pure consciousness, like numerous clouds precipitated within the same empty sky. The only stuff present in the clouds is the sky itself. The clouds, as such, give the sky a temporary name and form.

The reason we all seem to share the same world is not that there is one world 'out there' known by innumerable separate minds, but rather that each of our minds is precipitated within, informed by and a modulation of the *same* infinite consciousness. There is indeed one world that each of us shares, but that world is not made of matter; it is a vibration of mind, and all there is to mind is infinite, indivisible consciousness.

It is precisely because the world appears in and is made of infinite, indivisible consciousness that it appears from the perspective of each finite mind to be the same world. It *is* the same world, because all finite minds are refractions of the same consciousness. It is the sameness of consciousness, which shines in and as each of our minds, that is responsible for the conviction that we all share the same world. In the same way, all the characters in a night dream feel that they share the same world because they are all created by the same dreaming mind.

The sameness of the world is the sameness of consciousness. Our shared world is our shared consciousness. The vastness of the universe is the vastness of consciousness. Each finite mind feels that the world is much bigger than itself, and this intuition is true. There *is* more to the world than an individual mind, but this doesn't imply that the world is outside consciousness. When a mind experiences the vastness of the universe, it is experiencing a segment of God's infinite being from its own limited perspective, and it is for this reason that we feel such awe and wonder before nature.

Materialists use the intersubjective agreement – the agreement that individual minds share their experience of the same world – as proof that there is an independently existing world outside consciousness. However, that is just an interpretation. This intersubjective agreement can also be used to assert the opposite point of view, namely that the similarity of everybody's experience of the world is an inevitable consequence of the shared nature of our minds at their deepest level.

One might argue that it is not possible to choose between these two opposing assertions, both of which use the same evidence – our shared world – as their proof. However, there is a difference between them. The materialist perspective is not grounded in experience. It requires an abstract line of reasoning that presupposes the existence of a reality outside consciousness, although nobody has ever experienced this, nor could they ever experience it. The materialist point of view asserts the reality of that which is *never* experienced – matter – and denies that which alone is *always* experienced – consciousness itself. That is the tragedy and the absurdity of the materialist perspective from which humanity is suffering.

The second point of view – which is not just the spiritual but also the truly scientific point of view – is in line with our experience and so should trump the materialist perspective, which turns out to be nothing more than a belief and, as such, simply a popular religion. For this reason, the materialist perspective should be sliced out of our contemporary worldview with Occam's razor* and the laws of physics, as a result, should be upgraded to laws that govern the unfolding of *mind* rather than the behaviour of matter.

*A problem-solving principle attributed to William of Ockham (c. 1287–1347), an English philosopher and theologian. The principle can be interpreted as stating: 'Among competing hypotheses, the one with the fewest assumptions should be selected.'

For centuries our culture has been dominated by the materialist view of reality. It is not necessary to point out the devastating effects of this view: the extent of suffering and conflict in society speaks for itself. If the human race still exists in five hundred years' time, hopefully people will look back on this period of materialism just as we now look back on the theories of a flat earth and a geocentric universe that dominated our world culture for centuries. If humanity does not still exist in five hundred years' time, it will most likely be because materialism prevailed.

Humanity cannot survive the materialist paradigm. If our species, and countless others, are to survive, we will have to replace the matter model with the consciousness-only model. If we want to build a model of experience, we have to start on solid ground, that is, we have to start with experience. If we build a paradigm starting with a belief, that belief will inform every aspect of the paradigm, and everything that proceeds from it will simply be an expansion of the fundamental assumption contained within it.

Experience must be the ultimate test of reality, and therefore the ultimate science must be the science of experience itself. All there is to experience is mind, and all there is to mind is consciousness. Thus, the ultimate science must be the science of consciousness.

The science of consciousness is consciousness's knowledge of itself. Consciousness's knowledge of itself – which is the only knowledge that remains the same at all times, in all circumstances and under all conditions and is, therefore, absolute knowledge or truth – must be the foundation and fountain of all relative knowledge.

A culture that is based upon any other understanding is bound in the end to destroy itself, for the ignorance at its heart – the ignoring of reality – will sooner or later rise up and turn people against themselves, their planet and one another.

CHAPTER 16

THE MEMORY OF OUR ETERNITY

We normally believe that experience takes place in time, and 'now' is the name that we give to the moment in time at which experience occurs. Time is, as such, considered to be a line extending indefinitely into a past and a future, and the now is considered to be a point that is slowly moving along that line.

For instance, we believe that breakfast tomorrow will exist at some time in the future, and that we are slowly moving along a line of time towards that event. Likewise, we think that breakfast yesterday took place at a moment in the past that is separated from the now by an ever-increasing duration of time. If we go back even further, we believe that there was a moment in the distant past at which we were born, and that we are progressing along this everlasting line of time towards a moment in the future when we will disappear or die.

Let us subject this model to the scrutiny of experience, because almost everything we think and feel, and subsequently almost all our activities and relationships, are predicated upon this model of time. Let us first think of an event that happened in the past. Although that thought is a thought *of the past*, nevertheless it takes place *now*, in the present. So instead of *thinking* about the past, try now to actually *experience* it. In order to do so we have to take a step out of the now and actually visit the place called 'the past'. Try to actually go to the time at which breakfast yesterday took place, not the thought or the image of it – these both take place now – but the actual experience of it.

Now think of a time in the future when breakfast tomorrow will take place. The thought takes place now, but try to actually experience that time. In order to do so we have to leave the now and visit this place in the future. See, in this way, not just philosophically but experientially, that these two times, the past and the future, are never actually experienced. They are concepts that, although essential for practical purposes, don't bear any relation to actual experience.

Go to the experience of now. How long does the experience of now last? Even to ask the question as to how long the now lasts is to assume time, which can never be found in experience. Something that lasts in time must have a beginning and an ending. Is the experience of now something that started at one moment and will stop at another? Have you ever experienced the beginning of now? If so, when did it begin and what was present prior to it? In order to experience the beginning of anything we must be present prior to the appearance of that thing. That is, we can only claim that it had a beginning if we witnessed it coming into existence. Likewise, have you ever experienced the end of now, and if so, what came after it?

How many 'nows' have you ever experienced? For instance, how many nows have there been today? Are there numerous nows, or is it 'always' the same now? Did now come from a distant past, and is it progressing towards a distant future? If so, at what speed is it moving? If we have no actual experience of a past or a future, what legitimacy is there to the belief that now is a moment that moves from the past to the future? A moment has duration, and something that has duration lasts in time. But there is no time present in which now can last. Now is not a moment in time. This now is the only now there is, and it is not going anywhere.

See that it is always now. However, what does the word 'always' mean? It means lasting forever in time. To say that now is always present means that the now lasts forever, throughout the past and the future, but we never actually experience a past or a future. There is no such thing as a 'present moment', because there is no time present for the now to be a moment within. If we agree that experience must be the test of reality, we cannot legitimately say that it is *always* now, for there is no time in which now can last. It is not always now; it is *eternally* now.

Eternal does not mean everlasting in time. It means without beginning or end and, therefore, not in time at all. Indeed, there is no time present in which now could be or last. The eternal now is not an extraordinary, esoteric or metaphysical dimension that none of us have access to under normal circumstances. It is the very now at which all experience takes place – not the everlasting now, but this, the ordinary, only, ever-present now. Time is not the container of now. Now is the container of experience, including all thoughts of time, but never the actual experience of time. Now is eternity.

Notice that whenever it is now (and it is 'always' now), I am. In other words, the eternal now is the seat of consciousness. If we 'go to' the experience of now, and then 'go to' the experience of being aware, we 'go to' the same experience. Now *is* 'I am'. Whenever it is now, I am; whenever I am, it is now. Now *is* consciousness. Thought cannot know consciousness, although it is made of it, and as a result it superimposes its own limitation on consciousness and conceives it as time. Thus, time is what eternity looks like when refracted through the limitation of thought.

In other words, thought, being ignorant of consciousness, conceives time in its place and confers the reality that properly belongs to consciousness onto time. Time becomes the container of experience, and consciousness is reduced to an infinitesimal fragment of time called the now. Thus, the now is a trace in apparent time of the presence of consciousness. It is, as such, the portal through which the finite mind must pass on its way back to its reality, and the same portal through which consciousness passes as it brings time and its contents out of eternity into apparent existence. Thus, the contents of consciousness always appear in time but they emanate from and within eternity. It was in reference to this understanding that William Blake said, 'Eternity is in love with the productions of time.'

* * *

Reconsider now the three states of waking, dreaming and sleeping: they are not spread out sequentially, for there is no time present during which they can occur. The waking, dreaming and deep sleep states are always experienced now, in this, the only now there is. They are, as it were, piled up on top of one another, although even this formulation is a concession to a spatial dimension of experience which we will later collapse.

In deep sleep there is no finite mind – that is, no activity of thinking and perceiving – and therefore no experience of time or space. In deep sleep there is only the self-aware presence of consciousness simultaneously being and knowing its own inherently peaceful, unconditionally fulfilled being.

This dimensionless field vibrates within* itself and assumes the form of mind, that is, takes the shape of the dreaming and waking states. As such, it colours itself with its own activity, thereby giving its own dimensionless presence the appearance of dimensions, that is, the appearance of time and space.

Just as a movie and a documentary appear on a screen, giving the two-dimensional screen the appearance of three dimensions, but never actually displace or modify the screen itself, so the states of dreaming and waking emerge 'within' the ever-present, dimensionless field of pure consciousness but never displace or modify it. In fact, the deep sleep state is only considered to be a limited state from the point of view of either the waking or dreaming state.

Believing that time and space are attributes of an objective reality that exists independently of mind, thought imagines that in deep sleep time and space (and all the events and objects they contain) continue, and that mind and its essence – consciousness – disappear. This belief is an inevitable corollary to the belief that the body is made of matter and that mind is derived from it. From the perspective of the consciousness-only model, it is consciousness itself that assumes the form of mind, that is, thought and perception, each mode projecting its own limitations on consciousness and making it appear to itself as time and space.

From this perspective deep sleep is not a state that a person enters; it is simply consciousness divested of the activity of mind. As a concession to the belief in the existence of objects, we could say deep sleep is the state of consciousness without objects. In fact, a person never enters or passes through any state of waking, dreaming or deep sleep. There is no such person present to pass through any state. Every state of mind is a self-colouring of consciousness; at no point does any entity other than consciousness itself ever come into existence.

In other words, the mind superimposes its own limitations on consciousness and conceives of consciousness without an object as a state of deep sleep that takes place in time. However, there is no experience of time or space in deep sleep. And for good reason! Deep sleep does not take place in the time or space that seem to exist from the perspective of the waking and dreaming states.

---

*Of course, there is no 'within' to something that has no dimensions – indeed, 'something' that has no dimensions is not a thing – but there is no language for this understanding, so we must consent to use words that have evolved to describe the events and objects of time and space, and yet not be bound by their implicit limitations.

In fact, the reason we love to fall asleep is that in sleep, our essential being of pure consciousness is relieved of its activity of mind and knows its own inherently peaceful and unconditionally fulfilled being, the experience commonly known as peace or happiness.

* * *

All experience takes place in the eternal now, the only now there is, and this now is not going anywhere. This very now in which your current experience is taking place is the same now in which every experience you have ever had took place. Indeed, your birth took place in this now, although this now did not come into being with your birth, and your death 'will' take place now, although this now will not go out of existence when you die. As T. S. Eliot writes in *Four Quartets*:*

> Or say that the end precedes the beginning,
> And the end and the beginning were always there
> Before the beginning and after the end.
> And all is always now.

Imagine a novel that consists of the life story of a hundred-year-old woman. The time that seems to exist for the woman in the novel is present *simultaneously* in the form of the complete book. However, the mind only has access to the story one word at a time. Therefore, it is the mind that superimposes its own limitations on the novel and creates time out of what is, in fact, simultaneously present.

The time experienced is not in the book; it is in mind. It *is* mind! Mind superimposes its own limitations on eternity and makes eternity appear in a way that is consistent with those limitations. Time is created in mind; mind is not created in time. Time is, as such, what eternity looks like from the limited perspective of mind. It is consciousness itself that assumes the form of mind in order to bring time out of eternity and space out of infinity.

Consciousness is said to be both eternal and infinite as a concession to the twin modes of mind, thought and perception. Thoughts seem to appear in time, perceptions in space. From the perspective of thought and perception – that is, from the perspective of the finite mind – time and

*From 'Burnt Norton', *Four Quartets* (1935).

space are the containers within which experience takes place. As such, time and space are the two subtlest and most transparent objects. They mimic the form of consciousness in the finite mind by sharing its transparent qualities.

In order to manifest a world, the eternal, infinite nature of consciousness must be downgraded to time and space so that thoughts and perceptions may occur. It is thought and perception that bring eternity and infinity out of being into apparent existence. Eternity gives birth to time within itself; infinity unfolds space within itself. However, at no point does consciousness ever cease being dimensionless consciousness. From consciousness's pointless point of view it is eternal and infinite.

It is not possible to think about the now, because thought imposes its own limitations upon everything it conceives, therefore when mind thinks of the now it imagines a fraction of a moment sandwiched between the past and the future. For the same reason, it is not really possible to write coherently about these matters. We can only hope that in thinking and writing about these ideas a few words may trigger in the mind the memory of its eternity, and that the perfume of this memory will attract it in a new and directionless direction.

* * *

All experience is believed to take place in space, thus space is considered to be the medium in which mind is located. 'Here' is considered the particular point in space from which the rest of space and its contents are known or viewed. Everything that is not 'here' is considered to be 'there'. But do we ever actually experience the place called 'there'?

The experience of 'there' takes place *here*, just as the experience of the past or future takes place *now*. It is not possible to leave 'here' and visit 'there'. 'There' is always a concept, never an experience. Space is the distance between the point 'here' and the point 'there', or between two points 'there'. However, only *here* is experienced. How much distance can there be between that which is experienced and that which is not experienced? See in this way that space itself is a concept and never an experience.

Notice that wherever experience is known, it is here. Here is, as such, the 'place' at which experience takes place, but that place is not in space. Here is the place at which I always am. It is the place at which consciousness is.

Notice that whenever it is here, it is also now. The here and the now always intersect. In fact, they do not intersect; they are the same 'point'. They are not even a point, but we cannot conceptualise, let alone speak, of something without dimensions, so let us consent to consider the here and now an infinitesimally small point that is not located in the time or space in which experience takes place.

The landscape of which we dream at night does not take up any space in the dreamer's mind, and the lifetime that seems to occur in the dream takes only a moment of waking-state time. Likewise, the time and space that seem to exist in the waking state are eternal, infinite consciousness refracted through the prism of thought and perception. In reality, experience does not take place in time and space, but consciousness has to assume the appearance of time and space – it must collapse into the finite mind – in order for there to be experience.

The word 'infinite' does not mean extended indefinitely in all directions. Infinite means 'not finite', that is, with no finite qualities. It means without dimensions, that is, not in space, just as eternal means not in time. The infinite is not an extraordinary, mystical realm to which none of us have access. The infinite is the placeless place, dimensionless consciousness, at which all experience takes place; we are 'always' experiencing it. The infinite refracted through the prism of perception appears as space, just as eternity refracted through the prism of thought appears as time. It is the limitations of our own minds that make eternal, infinite awareness appear as time and space.

*   *   *

We normally believe and feel that we pass through three states of waking, dreaming and sleeping each period of twenty-four hours. The only reason we believe and feel that we pass through these states is that we have forgotten, overlooked or ignored our essential nature of eternal, infinite awareness. Eternal, infinite awareness never passes through any state; all states pass through it. In fact, we have not forgotten, overlooked or ignored our essential nature of eternal, infinite awareness, for we are not something other than awareness that can know or forget awareness. We *are* awareness itself. It is awareness itself that seems to have veiled itself by assuming the form of mind, thereby colouring itself with its own activity and appearing to itself as objective experience.

In fact, it is not quite right to say that all states pass through awareness, although to do so is a legitimate concession, in the early stages of this exploration, to the mind that believes in the independent existence of objects and states. To say that all states pass through awareness suggests that a state comes from somewhere outside awareness, passes through awareness and then leaves, like a train passing through a station. There is nowhere outside awareness from where an experience or state could come, nor anywhere outside awareness into which it could pass.

A movie doesn't *pass through* the screen; it *is* the screen. All there is to a movie is the screen. The movie is a self-modulation of the screen. The three states of waking, dreaming and sleeping don't pass through eternal, infinite awareness; they are self-colourings of eternal, infinite awareness. In fact, only the waking and dreaming states are colourings of awareness. It is only from the perspective of the waking and dreaming states that deep sleep is considered to be a third state that exists in the time and space that are, from the perspective of these two states, considered real in their own right.

Imagine two teenage boys sitting on the sofa watching television. Their mother comes in and asks the first son, 'What are you watching?' He says, *'Breaking Bad!'* An hour later she comes in and asks the first son, 'What are you watching?' This time he says, *'The Simpsons!'* An hour later she comes in. The show is over and the TV is turned off, and the mother asks the first son, 'What are you watching?' He says, 'Nothing!'

The mother also asks the second son the same series of questions. She comes in during *Breaking Bad* and says, 'What are you watching?' He says, 'I see the screen!' An hour later she comes in again: 'What are you watching?' 'I see the screen!' She comes in an hour later and the TV has been turned off: 'What are you watching?' 'I see the screen!'

The first boy thinks that he is seeing a multiplicity and diversity of objects during *Breaking Bad* and *The Simpsons*, and therefore when the programmes end he thinks he is seeing a third state of the screen: nothing. The 'nothing' that he thinks he sees at the end of the two programmes is only in relation to the apparent multiplicity and diversity of things that he thought he was seeing during them. The second son doesn't superimpose *something* onto the screen during the two programmes, and therefore doesn't feel that he is seeing *nothing* when the TV is turned off. He sees the same screen throughout.

If we think that the waking and dreaming states comprise a multiplicity and diversity of separate objects and selves, we will consider deep sleep a blank, empty nothing. But if we understand and feel during the waking state that what appears to be a multiplicity and diversity of objects is in fact the single, indivisible, eternal, infinite screen of awareness, then we will no longer superimpose 'nothing' onto deep sleep. Deep sleep will be experienced as the 'uncolouring' but not the absence of awareness. The waking and dreaming states are a self-colouring of awareness; in deep sleep awareness remains wide awake but ceases to colour itself with the activity of mind. As our experience of the 'somethingness' of the waking state dissolves, so the 'nothingness' of deep sleep subsides in proportion.

Deep sleep is not a new experience that we have upon falling asleep. It is simply the revelation of the ever-present background of all experience. It is the uncolouring of awareness. The peace that is experienced in deep sleep is not something that is just available to us for three or four hours at night. It is continuously available. Any of us could fall asleep now and, as a result, experience the peace of our true nature. Even if we were deeply depressed now, or in emotional turmoil, all of that would vanish if we fell asleep, and we would instantly experience the peace and fulfilment of our true nature.

That peace and fulfilment is not a new experience that comes to us when we fall asleep. It is just the revelation of the peace that is ever-present in the background of all experience but usually obscured by the clamour of objective experience. Meditation is the art of falling asleep whilst remaining awake, thereby accessing the peace that is eternally present in the background of all experience, irrespective of its content. It is not necessary to turn the movie off to see the screen.

*   *   *

We never go anywhere. We do not pass through experiences; experiences are a self-colouring of the eternal, infinite presence of awareness. This presence of awareness is never stained or harmed or moved by experience. It is always in the same pristine, luminous condition.

There are states of mind, but no states of awareness or consciousness. Awareness is always in the same condition. The so-called states of waking and dreaming are self-colourings of the ever-present, inherently peaceful presence of awareness. This inherently peaceful presence of awareness is available equally to all people, at all times, under all circumstances.

Of course, that is just a manner of speaking. Awareness is not available to or known by people. Awareness is available to itself at all times, although it sometimes veils itself with its own activity.

Awareness does not fall asleep at night. Awareness knows nothing of sleep; it is always wide awake. In sleep, or indeed in any other state, awareness does not enter a new state. Awareness never passes through any state. What thought considers to be sleep is simply the removal of thoughts, feelings, sensations and perceptions from the ever-present, wide-awake, inherently peaceful, unconditionally fulfilled background of awareness.

Thought considers sleep to be the absence of awareness, but in fact it is the awareness of absence. Awareness is always in the same pristine condition; nothing ever happens to it. In sleep awareness ceases to colour itself as the activity of mind and reveals itself to itself, and as a result we taste the peace of our true nature. Awareness recognises itself. That's why we look forward to sleep!

Awareness, the reality of all experience, is always in the same indivisible, indestructible, luminous condition. Understand and feel that nothing ever happens to you. You are still in the same pristine condition that you were in on the day you were born, and the day before your birth, because you, awareness, were never born. You are in the same pristine condition now that you will be in on the day after you die, because you, awareness, never die.

With this understanding, we no longer feel we have to defend ourself against any experience or retreat from life in any way. We are safe wherever we go. We are untouched by experience, and yet we touch all experience intimately.

*   *   *

We never go anywhere. A flow of images, sensations and perceptions moves through us; we do not move through them. We are not moving through time and space. Time and space are, as it were, moving through us.

When we travel, we experience a series of thoughts, sensations and perceptions, but the field of awareness in which these thoughts, sensations and perceptions move or, more accurately, of which they are a self-colouring, never goes anywhere. The common name for this field is 'here and now'.

Filtered through thought, the now of this field seems to be a moment in time. Filtered through the prism of perception, the here of this field seems to be a place in space. But the now that seems to be a moment in time, and the here that seems to be a place in space, are windows onto eternity and infinity. The true now, the only now, is the eternal now. And the true here, the place in which experience takes place, is the dimensionless presence of awareness. It never goes anywhere.

One of the reasons we love to travel is that that element of ourself which never goes anywhere is, by contrast with the movement involved in travelling, brought into focus. The changes emphasise the changeless. The sense of deep peace that often accompanies a journey is the peace of our changeless presence. We love to go to different places only to taste again and again the recognition that we never go anywhere and to feel the peace that accompanies that recognition.

I never go anywhere. I am always in the same place of 'I am', the placeless place called here, and the timeless time called now. Just as our essential being of ever-present, unlimited awareness shines in the mind as the knowledge 'I am', that knowledge appears in time as now and in space as here. And just as the knowledge 'I am' is a portal through which the finite mind or separate self must seem to pass on its way back to its essential, irreducible reality of pure consciousness, so the here and now is a beacon that shines in the midst of all experience, a secret door through which the mind passes out of time into eternity and out of space into infinity.

The screen never takes the journey that the character in the movie undertakes, although all there is to the journey is the screen. All there is to the appearance of space and time is infinite, eternal consciousness, but eternal, infinite consciousness never appears in space and time. Here and now is the intersection of infinite, eternal consciousness with the finite mind.

It is for this reason that as we get older we feel increasingly that we are not ageing. The belief that we are getting older seems to contradict our experience of always being the same person. As death approaches, we increasingly feel, 'How strange. I feel I am always the same person. I'm the same person I was when I was a five-year-old girl or boy. I'm not really getting older. What I was then is what I am now.' This intuition is one of numerous ways that truth intrudes upon the mind, infusing it with a trace of reality, the memory of our eternity, although thought will, in most cases, almost immediately dismiss it.

As Balyani said, 'He is now as He was then.' Our essential, self-aware being is always in the same pristine condition. At some point this intuition becomes our lived and felt experience, and it brings with it a release from the fear of death and the sense of lack that characterise the separate self or finite mind. Thus it brings with it the peace and fulfilment that are inherent in the knowing of our own being as it is, its knowing of itself in us, as us.

What I essentially am is eternally present. I didn't come from somewhere. I am not going anywhere. Nothing is ever added to me, nor removed from me. No experience ever aggrandises or diminishes me. I am complete, fulfilled, whole, perfect. What I essentially am is always in the same pristine condition. I have never been harmed, stained or aged by experience. I was not born and do not die.

The well of peace that rises from this understanding is the peace that passeth understanding. It does not come from the mind. It comes from the background of the mind but progressively floods the mind with its presence, gradually turning the mind into itself.

*   *   *

Just as now is the same eternal 'moment' for all people, which thought subsequently expands or unfolds as time, could it be that here is the same 'place' for all people, only seemingly diversified into a multiplicity of separate places in space by the activity of perception?

Imagine that you were to draw your current experience on a piece of paper. Remember that all you know is experience, and all experience is mind: thinking, feeling, imagining, remembering, hearing, seeing, touching, and so on.

The background and essence of mind is transparent awareness, so the piece of paper on which you draw is transparent, like plastic. Visualise a sheet of transparent plastic paper, and draw on it your current experience. Imagine that somehow you could make marks that would convey the current experience of thinking, feeling, sensing, seeing, hearing, imagining, tasting, smelling and remembering: your drawing is like a mini Jackson Pollock.

Now put that drawing aside and take out another transparent sheet. Draw the experience that happened yesterday and put it down next to the first sheet: another mini Jackson Pollock. Take out a third sheet, draw the experience that happened the day before and place it next to the second sheet. And the day before, and the day before, and the day before...and spread out these sheets in a long line, each sheet representing a day in your life.

Each sheet has a drawing of the content of your experience during that day, and they are laid down in sequence next to one another, representing the line of time of your life. The drawings representing your earliest experiences have just a few lines and dots floating around on the transparent page, and they become more complex as you get older. Sometimes they are dark and dense, and at other times they are more colourful and light.

Notice that at the time that each of the experiences represented in these drawings was taking place, it was now, and that we only ever experience one now. Therefore, it's not appropriate to lay the drawings out in sequence. So take all of your drawings and put them one on top of the other, creating a pile of drawings, each one on a transparent page, representing each day – or, we could say, each moment – of your life. Every experience takes place in the *same* now, and so the drawings are not laid out in sequence but rather stacked one on top of the other in parallel, as in a book.

It is true that, due to the limitations of mind, in order to access the story of our lives that is simultaneously present in the eternal now, we have to read one page at a time. Thus it seems from the perspective of mind that the events of our lives take place in time. In fact, it is the mind that unfolds the eternal now and makes it appear as time.

Now imagine you are sitting in a room full of people, each person sitting on a chair. Notice that the body is experienced in the mind – that is, it is an experience of sensing and perceiving – and is therefore part of the book of drawings. So visualise that there are not bodies sitting on chairs, but rather that on each chair sits a book, each book consisting of numerous transparent sheets, each sheet with a drawing on it representing a moment in the life of the person.

Notice that the sheet on which each of our drawings is made is transparent, empty, therefore there is nothing to distinguish one sheet from another. Although the drawings are numerous and colourful, the sheets cannot be distinguished from each other. The background of awareness is always the same indistinguishable, indivisible, space-like, self-aware presence in which all experience appears.

So now collapse your book so that its entire content takes place on a single transparent sheet, each drawing superimposed one upon the other, forming a dense and layered image that contains every experience that you've ever had in your life. There are now numerous transparent sheets, one per chair, each one representing the entire contents of each person's life.

In reality the chairs, the room, the building and the world itself are all also made of mind and should, therefore, be part of the drawings on the transparent pages, but I do not want to stretch the analogy to breaking point. So let us stay with the image of numerous transparent sheets, each one on the chair on which each person is sitting.

Notice that the now in which each person's experience takes place is the same now. So take the numerous sheets of paper from the chairs and place them on top of one another, representing that in fact the lives that the drawings depict all take place in the same eternal now. So we now have a single book again, only this time each page of the book is the entire contents of a single person's life, just as we previously had a book for each individual's life.

The transparent pages of this book are empty and without inherent objectivity, and therefore there is nothing to distinguish the transparent pages on which each person's life is drawn. So collapse the book again: each person's life takes place in the same now, so everyone's drawings are on the same sheet of paper. We now have a single sheet of transparent paper on which the entire contents of everyone's lives are drawn.

Notice that whenever it is now, it is also here, that is, the now and the here always coincide. Could you ever feel that it is now and not at the same time feel it is here? So the eternal moment at which this single sheet is positioned is also the same place 'here' at which everyone's experiences takes place. Now we only have to make one more step.

The entire contents of each person's life are drawn on a single sheet of transparent paper, which is a representation of the here and now. However, here is not a place in space and now is not a moment in time, and therefore neither has dimension. So in your imagination take this transparent sheet on which the entire contents of everyone's lives are drawn, and shrink it until it becomes a minute point.

I'm not going to go further than 'a minute point', because we cannot imagine or think about something without dimensions. But please understand that this minute point represents dimensionless consciousness, in which the entire contents of each person's life are condensed. This point is the 'here and now', the placeless place and the timeless time 'in' which and 'at' which experience takes place.

What is it that makes this unlocated, timeless 'point', in which all experience is contained, expand and seem to appear as time and space, in which an apparent multiplicity and diversity of objects and others seems to exist? It is the activity of mind: thought and perception. In order for the content of this minute point to become knowable, it must be known through the agency of a mind. Thought and perception are mind; through thought this point is expanded and appears as time, and through perception this point is amplified as space. Time and space emerge out of this dimensionless point through the agency of thought and perception.

In this dimensionless presence of awareness, all of our experiences are merged together indistinguishably, but as thought and perception expand out of this minute point, so diversity begins to grow and ramify, giving birth to itself as it expands like a branching tree, each branch creating more diversity, more complexity and, therefore, apparent separation.

The result of this process is the experience that each of us is having now. Every person and animal has emerged from this condensed essence of a single dimensionless consciousness. Each of us is, as such, the *same person*, apparently diversified and separated through the kaleidoscope of thought and perception. Just us, relatively speaking, each of our finite minds appears in the form of innumerable thoughts and perceptions without ever ceasing to be an integral whole, so every person is a thought in the mind of infinite consciousness, whose infinite and indivisible nature is never compromised by their apparent multiplicity and diversity.

The conventional interpretation of reincarnation makes the mistake of assuming the existence of time, and yet this assumption is based on a valid intuition. We do not reincarnate in time, and yet each of our minds, and all the minds of those who have lived before us, is born from a single mind, of which it is an extension. We are literally each other. Each of us is the outer face, or the objectification, of the only mind there is, eternal, infinite consciousness. We are all mirrors of the same consciousness.

As Primo Levi wrote:*

> ...remember the time
> Before the wax hardened,
> When everyone was like a seal.
> Each of us bears the imprint
> Of a friend met along the way;
> In each the trace of each.

*From 'To My Friends', *The Mirror Maker*, Schocken Books (1998).

CHAPTER 17

CONSCIOUSNESS'S DREAM

Imagine a woman named Mary who lives in New York. One night Mary falls asleep and dreams that she is Jane walking the streets of London. In this analogy Mary's mind stands for consciousness. Relatively speaking, Mary's mind is infinite, that is, all possible dreams lie in potential within it. She could dream she is Sophie in Amsterdam, Katie in Munich or Chloe in Rome, but tonight Mary dreams she is Jane in London.

Asleep in her bed in New York, Mary cannot visit London. Although Mary's mind is by nature infinite, she must freely allow her unlimited mind to assume the form of Jane's limited mind in order to simultaneously manifest and know the streets of London, thereby realising one of the possibilities that exist within her.

Amsterdam, Berlin, Tokyo, Vienna and Rome all exist in potential in Mary's mind, but to experience one of those cities she has to give up all the other possibilities. Her infinite mind has to collapse into Jane's finite mind, for it is only through and as Jane's finite mind that Mary can know the multiplicity and diversity of London. So, for Mary, manifesting the streets of London is a kind of sacrifice. She has to forget or overlook the knowing of her own unlimited mind as it is and freely consent to limit herself in order to 'become' Jane.

From Mary's perspective, both Jane and the streets of London take place within her own mind. However, from Jane's point of view, her own thoughts and feelings take place *inside* her mind, and the world that she sees – the streets of London – takes place *outside* her mind.

In order to know the streets of London, Mary has to dream the world within herself and then become a person *in that world*, from whose point of view she is now able to view it. In the same way, consciousness precipitates the world within itself and, at the same time, becomes a separate subject of experience *in that world*, from whose point of view it now seems to see or know it.

The infinite cannot know something that is finite; therefore, infinite consciousness cannot by itself know a multiplicity and diversity of objects, others or the world. Infinite consciousness needs an agency, a mechanism through which it can know or experience a multiplicity and diversity of objects, just as Mary needs the agency of Jane's mind in order to perceive the streets of London. Infinite consciousness needs the agency of the separate self or finite mind to actualise in form the formless potential that lies within itself.

The difference between Mary, asleep in New York, and consciousness is that Mary can have only one dream at a time – she has to be Sophie in Amsterdam, Katie in Munich or Chloe in Rome – whereas consciousness can have innumerable dreams at the same time. Each of our minds is a dream in the infinite mind of consciousness. As such, each of our minds is the agency through which consciousness realises a segment of its infinite potential.

* * *

Mary's mind is always an indivisible whole. None of the characters or objects that make up her dream ever actually divide her single mind into a multiplicity and diversity of objects and characters. Likewise, all there is to each of our finite minds is infinite consciousness, and nothing that takes place within the dream of experience in each of our minds ever divides infinite, indivisible consciousness into a duality of objects and selves.

However, Jane experiences a multiplicity and diversity of objects and others, all of which seem to take place outside of herself. Based on the evidence of her senses Jane thinks, 'The knowing or the consciousness with which I know the world lives in my brain.' She notices that when she closes her eyes the streets of London disappear and when she opens them they reappear, and she reasonably concludes that whatever it is that is seeing the streets of London must be located just behind her eyes. Likewise, when she blocks her ears the sound of traffic disappears, and when she opens them it reappears, leading her to believe that whatever it is that is hearing the traffic is located behind her ears. Collating the information from her senses in this way, Jane concludes that her mind is located in her head.

Continuing this line of reasoning, Jane observes that her feelings and bodily sensations are private and belong to her alone, giving rise to the belief and feeling that her essential identity is an amalgam of her mind and body. She concludes that whatever happens to the body happens to her. When it grows old and sick, she feels that she grows old and sick; when it dies and disappears, she believes she will die and disappear.

Jane believes that the knowing with which she knows her experience is located in and shares the limits and destiny of her body. However, the knowing with which Jane knows her experience doesn't live *inside her body*. It lives in Mary's mind in New York! In the same way, the knowing with which each of us is knowing our current experience doesn't live inside our bodies. It lives in consciousness – in fact, not 'in consciousness', for there is no inside of 'something' that has no dimensions. It *is* unlocated, unlimited consciousness! And just as Mary's mind does not live in the time and space in which Jane's experience seems to appear, so consciousness does not live in, nor is it subject to the limitations of, the time and space which seem to be real from the perspective of the waking-state mind.

Just as Mary has to fall asleep to, and thus ignore, the reality of her own infinite mind in order to manifest and know the streets of London, so consciousness has to fall asleep to the reality of its own infinite mind to assume the form of each of our finite minds, from whose point of view it can know finite experience. The world, conceived by the finite mind as a multiplicity and diversity of objects made out of dead, inert stuff called 'matter', comes into apparent existence when consciousness ignores the reality of itself, and it vanishes out of apparent existence when consciousness wakes up to or recognises itself. When Mary falls asleep, Jane wakes up, and when Mary wakes up, Jane falls asleep.

In order to know Jane's finite experience Mary must overlook, forget or ignore the infinite nature of her own mind, although all Mary ever knows, even in the form of Jane's mind, is her own indivisible mind. Similarly, consciousness must seem to overlook, forget or ignore its own nature in order to know the world, even though its experience of the world is in fact its experience of itself, albeit seen through the lens of the finite mind.

However, that is not how things seem to be from Jane's point of view. Although in reality – that is, from Mary's point of view – there is only Mary's own indivisible mind, from Jane's point of view experience seems to be divided into a multiplicity and diversity of separate objects and selves which exist outside her mind, and thoughts and feelings which exist inside her mind. Jane superimposes the limitations of what she considers to be her own mind on her experience and sees everything through its distorting lens. Likewise, the finite mind superimposes its own limitations on reality and sees in it a reflection of those limitations, which it mistakes for reality itself, whilst all along, all there is to reality is the infinite, indivisible reality of consciousness itself.

When I say that the mind superimposes its own limitations on reality, I do not mean to suggest that these limitations come from outside consciousness. There is no such place! It is consciousness itself that freely assumes the form of the finite mind in order to bring manifestation out of being into existence. Thus, although the mind seems to fragment and diversify reality, just as an image seems to fragment and diversify the screen on which it appears, it is itself a manifestation of that very reality. We could say, therefore, that reality veils itself with its own activity and at the same time fully expresses itself *as* that activity.

'Mind' is the name Jane gives to the stuff out of which her *subjective* experience – her thoughts and feelings – is made. 'Matter' is the name she gives to the stuff out of which her *objective* experience – the world, objects and others – is made. Thus, from Jane's point of view, her experience is divided into two substances, mind on the inside and matter on the outside.

Moreover, from Jane's point of view, her body, and the mind that she believes resides inside it, are obviously the product of the world in which she lives. For this reason it seems quite reasonable to her to conclude that her mind is a by-product of matter. The evidence of her senses seems to support this belief, and most of the people in her dream world corroborate her experience. She overlooks the fact that a brain is an experience in the mind – a thought, sensation or perception – and thus misses an important hint as to the nature of her mind and the world it seems to experience.

In the same way, from the point of view of the separate self or ego, experience seems to be divided into two essential substances: matter is considered to be everything that exists outside of mind, from which the mind itself is considered to be derived. In other words, matter is considered to be primary and mind secondary, in contradiction with our experience.

The fact that all that is or could ever be known of the world or the body are appearances in the mind, and that the only substance present in mind is consciousness, is a vital clue for the mind as to the essential nature of its experience. However, so hypnotised has the mind become by its own creativity, and so profound is the ensuing amnesia, that it misses that clue and repeatedly asserts the existence of a substance called matter that it has *never* experienced in favour of the sole reality of its entire experience, consciousness itself.

*　　*　　*

Who is Jane? Jane is not a self, a person or an entity in her own right. She is an apparent limitation of Mary's mind, but apparent only from her own illusory and ultimately unreal point of view. From Jane's point of view she is a self, a person, an entity, a body. But that is simply a belief! Jane thinks that she *has* thoughts, whereas in fact she *is* a thought! She is a thought and a subsequent feeling that impose a limit on the true identity of Mary.

This does not mean that Jane is unreal or non-existent. It simply means that she is an apparent but unreal limitation on the only 'one' that truly is: Mary. From Mary's point of view, Jane is not a separate self or person. From Mary's point of view there are no objects or selves, no real *things* in her dream, no entities with their own independent existence that can be defined by nouns. There is only the *activity* of her own mind, which is best described with verbs. In the new language of non-duality there are no nouns. Everything is understood as the activity of consciousness.

The apparent self of Jane is the true and only self of Mary. What Jane considers to be 'I' is simply an imaginary limitation of the true and only 'I' of Mary. Jane's apparently finite self or being is Mary's infinite being – the only being there truly is – with an imaginary limit attached to it. But that limit is not a mistake, unless it gives rise to the belief and feeling of a separate and independently existing entity.

That separate and independently existing entity, called 'I', is unreal in the sense that it never actually exists in its own right. However, like all illusions, it has a reality to it. The reality of the water in a mirage is light, just as the reality of the landscape in a movie is the screen. The reality of Jane's finite mind is Mary's infinite mind. Jane is a *process*, not an entity. She is the agency through which Mary is able to manifest and know the world.

The ego, separate self or finite mind is the process through which consciousness manifests a segment of its infinite potential. As a process, the ego is not a mistake or a problem; it is the mechanism of creation. But as an entity it *is* a problem. The belief that we are separate, independent entities is the fundamental assumption upon which our entire world culture is founded.

This separate self or ego believes and feels itself to be a fragment. Being a fragment, it always feels that it is incomplete and therefore in need of fulfilment; that it is vulnerable and therefore in need of protection. The separate self is always unhappy and afraid, and all of its activities and relationships are designed to alleviate the discomfort of this unease or suffering. Thus, the belief in separation is the ultimate cause of the unhappiness we experience within ourselves and the conflicts we experience between individuals, communities and nations.

*　　*　　*

Jane notices that everyone on the streets of London seems to be experiencing the same world she does, and she concludes from this that everyone is seeing a partial view of the same world. At the same time, she notices that whilst everyone seems to have access to the same world, they also have their own private thoughts and feelings which apparently take place inside their minds, in each of their heads.

Jane does not realise that what she considers to be consensus reality, the world she shares with all other people, is a reflection of the limitations of her own mind. Likewise, the world that each of our finite minds knows and experiences reflects the limitations of the mind through which it is known. However, just as there is an undoubtable reality to Jane's world – Mary's mind – so there is a reality to the world we perceive in the waking state. The world that we perceive is real, but its reality is infinite consciousness. It is infinite consciousness, filtered through the limitations of the mind, that appears to itself as the world. In religious language, the world is what God's infinite mind looks like from the perspective of a separate subject of experience. Matter is consciousness refracted through the prism of the finite mind.

Let's imagine now that in Mary's dream, Jane invites Peter, Clare and John for dinner. Mary's mind takes the shape of the entire dinner party: the house, the room, the view, the food, and Jane, Peter, Clare and John. The four friends start discussing their experience. There is a bowl of fruit in the middle of the table, and they compare notes about it. They all report that there are six apples and eight oranges in a white bowl. They agree that although each of their minds is obviously separate and distinct from the others – none of them knows what the others are thinking or feeling – they all see the same bowl of fruit and can accurately describe and corroborate their experience. This they consider sufficient proof that a shared world made of matter exists outside consciousness.

Why do Jane, Peter, Clare and John see the same object? Because the object exists in its own right, outside and independent of consciousness? No! It is because each of their finite minds is precipitated within, informed by and made of Mary's *single* mind. It is the singleness of Mary's mind – the source from which each of their apparently separate minds proceeds – that enables each mind to see the same world. Likewise, the reason we all see the same world is due to the fact that each of our finite minds is precipitated within and informed by the same consciousness. It is the sameness of consciousness that gives rise to the apparent sameness of the world.

Jane, Peter, Clare and John don't realise that the knowing with which they know their experience belongs to Mary. It *is* Mary! They all know their experience with the *same* knowing. The bowl of fruit lies in potential in Mary's mind. Each of their finite minds brings into existence a partial view of this unmanifest potential which, from their limited perspectives, appears as a single object in their shared world. It is true that in Mary's dream Peter, Clare and John don't have their own subjective experience – only Jane does – but I am stretching the metaphor to represent the normal waking state, in which it is reasonable to infer that the other people we encounter have their own experience.

In fact, just as the multiplicity and diversity of thoughts and images that appear in Jane's finite mind never fragment her mind or divide its essential homogeneity, so the apparent multiplicity and diversity of finite minds that appear in infinite consciousness never divide infinite consciousness into numerous minds or compromise its unicity. Separate minds are only such from the limited and ultimately illusory perspective of one of those minds.

*   *   *

The singleness of the bowl of fruit that appears in Jane, Peter, Clare and John's shared world is not a proof of a world outside Mary's mind, in just the same way that the world that each of us now sees is not a proof of the existence of a world outside consciousness, made of matter. On the contrary, the same evidence can equally be used as an indication that the world we share, and that we view through the various facets of our finite minds, has a single source in infinite consciousness. How to choose between these two possibilities? Both Occam's razor and the evidence of experience support the consciousness-only model.

When Jane says, 'I see the bowl of fruit', she refers to the 'I' that sees or knows, and she thinks this 'I' belongs to and resides in her body. But the 'I' in Peter, the 'I' in Clare and the 'I' in John – the knowing with which each of them knows their experience – is the same 'I', the same knowing, *Mary's* knowing. It is this single knowing 'I' that has refracted itself in the form of each of their four finite minds.

If Jane, Peter, Clare and John were to take the thought 'I' and trace it back, asking themselves the question, 'What is the fundamental nature of the knowing with which I know my experience?', each of their minds would begin to travel backwards or inwards towards the source of its knowing. In doing so their minds would gradually be divested of their limitations and, at some point, be revealed as Mary's single, indivisible, infinite mind. Just as a slowly fading image is revealed to be made only of the screen, so the mind that investigates its own nature is revealed to be made only of infinite consciousness. The finite mind does not *become* infinite consciousness. It is relieved of its limitations and, as a result, stands revealed as the infinite consciousness that it always only is.

Only a finite mind can know a finite object, other or world; only infinite consciousness can know infinite consciousness. So if Jane, Peter, Clare and John want to know the nature of their own minds, each mind must seek its source, thereby divesting itself of all limitations. However, Jane, Peter, Clare and John are not separate entities that *have* minds. There is just one infinite mind – Mary's mind – that has the capacity to precipitate within itself several finite minds, each of which views the world – Mary's mind – from its own limited point of view. In fact, each of their minds does not *have* a point of view; it *is* a point of view. It is a temporary localisation through which and as which Mary's infinite mind is able to know itself as the dinner party, just as each of our finite minds is a point of view through which and as which consciousness knows itself as the world.

Infinite consciousness has the ability to precipitate within itself innumerable temporary, local points of view, each of which corresponds to an individual, finite mind. The structure or configuration of each mind will condition its experience: if infinite consciousness takes the shape of a human mind, such a mind will perceive, or rather take the shape of, the world that we all know. If it takes the shape or configuration of a dog's mind, a whale's mind, a sparrow's mind or a spider's mind, it will see or appear as the corresponding world. But the world that is seen in each case will not be something that is outside of consciousness. Nor, indeed, is it *in* consciousness. It *is* consciousness!

It is true that the world that Jane, Peter, Clare and John perceive is outside their finite minds, but that world is not outside Mary's mind. In fact, the 'outside' world that each perceives is made out of exactly the same seamless, intimate stuff of which their 'inside' thoughts and feelings are made. Although, from the four friends' point of view, experience seems to be divided into mind on the inside and matter on the outside, the former apparently derived from the latter, in reality both their interior experience and their exterior experience are made out of the same infinite, indivisible substance. There is only one reality – Mary's mind – and all apparent things are modulations of that.

There is only consciousness, modulating itself in all forms of experience. Each world is a facet or cross-section of consciousness's own infinite potential, the indivisible reality of all experience. The localisation of a temporary mind or temporary point of view actualises part of that potential.

Let us imagine that in Mary's dream Jane and Peter are partners, so when they look at each other across the table they feel great love for one another. What is that feeling? It is the intuition that at the deepest level they are one. Love is the experience of our shared being filtering into the finite mind. Jane and Peter are right: the two finite minds with which each of their experience is known, and out of which it is made, *does* indeed have a common source. Both minds are facets or localisations of Mary's single mind.

Their two minds are, in fact, never separate. The apparent separation between minds is only real from the limited perspective of one of those minds. It is the indivisibility of infinite consciousness that is responsible not only for the four friends' agreement about their perception of the bowl of fruit, but also for the experience of love or friendship. Just as our shared world is evidence of the unicity of consciousness from the outside, so love is its evidence on the inside.

In religious language, the feeling of love is God's footprint in the heart. It is the experience of our shared being. Likewise, the thought 'I am' is God's signature in the mind. The knowledge 'I am' is the shared light of infinite, indivisible consciousness refracted into an apparent multiplicity and diversity of selves or minds. It is for this reason that the spiritual traditions elaborate two paths, the paths of love and knowledge. Through the path of love or devotion, all feelings are traced back to their source of love; through the path of knowledge, all experience is traced back to its source in infinite consciousness. The two paths are gateways to the same feeling-understanding.

*   *   *

Everyday life is full of experiences that halt the normal workings of the mind and, in that brief moment, give the mind access to its own reality. Moments of intense fear, wonder, awe, heartbreak, joy, sorrow, love are all experiences that at least potentially have the power to cut through or dissolve the normal subject–object relationship and reveal the deeper shared reality that is common to both.

However, there is a gentler way that this recognition may be evoked, and it is available to everyone under all circumstances. It is simply to ask ourself the question, 'Who or what is it that is aware of my experience?' With this question, the knowing which is normally directed away from itself towards the objects of experience – thoughts, feelings, sensations and perceptions – is directed back towards itself. The mind is investigating its own essential nature, rather than its contents. This is meditation, self-enquiry or prayer. It is as if Jane were to ask herself, 'What is the nature of the knowing with which I know my experience?'

If the mind explores its own reality in this way, it recognises that its essential, irreducible, unconditioned nature is simply being aware or awareness itself. 'I' is the name the mind gives to this simple experience of being aware or awareness itself, and as a result of this recognition we begin to feel 'I am awareness' as strongly as we used to feel 'I am the body'. The mind's essential, irreducible, unconditioned nature is revealed as awareness itself. Jane's finite mind is revealed as Mary's infinite mind.

From the point of view of the waking-state mind, consciousness is a phenomenon that seems to appear and disappear. But from the point of view of consciousness itself – and consciousness is the only one that knows consciousness – it experiences itself as ever-present and without limits. It has no experience of its own appearance, evolution or disappearance. It never experiences itself being born, growing old or dying. However, from Jane's point of view her mind seems to be contained within her body and to share its limits and destiny. Along with everyone else in her world, she is completely hypnotised by the limitations that her own mind superimposes upon her experience, and were it not for a single occurrence, she would never question this.

However, Jane is suffering, and her unhappiness causes her to strive ceaselessly for happiness. In fact, her entire life is orientated towards the relief of her unhappiness through the acquisition of objects, substances, activities, states of mind and relationships. She goes from one object or relationship to another, hoping that they will put an end to her suffering. Indeed, when she acquires the object, meets in friendship or merges in sexual intimacy, her suffering does come briefly to an end. As a result Jane believes that if she could just acquire more objects or have more relationships she would become happier and happier.

Jane spends more and more time seeking fulfilment in objects, substances and relationships, but nothing works. Although she intuits that none of these objects or relationships will ever truly fulfil her, she does not know where else to turn, so profoundly has she been conditioned by her culture to believe that happiness is derived from objects.

However, this intuition is enough to open Jane's mind to another possibility, and one day, in response to this openness, Jane meets her friend Clare in a café. Clare says to her, 'Jane, your suffering is not caused by a lack of objects and relationships. It is due to the fact that you have forgotten or overlooked what you essentially are. You have lost yourself in experience.' Jane looks puzzled and replies, 'But I am suffering terribly.'

'What do mean by the word "I"?' Clare asks. To begin with, Jane describes her thoughts, feelings and sensations, but Clare points out that these are continuously coming and going and cannot therefore be what she essentially is. 'What you essentially are must always be with you', she says. 'What element of yourself remains present throughout all your changing experience?'

Jane closes her eyes, and she begins to feel her attention sinking deeper and deeper into herself, through all the layers of changing thoughts and feelings, through all the sensations and perceptions that make up her experience of the body, ignoring anything that comes and goes until she cannot go any further. She remains silent for some time. 'I just am', she says at last.

'And how do you know that you are?' Clare asks. After another long silence Jane answers, 'It's self-evident. I *know* that I am because I am *aware* that I am.' Clare is reluctant to say anything. She knows that Jane is having the most profound experience it is possible to have, the simple knowing of her own being, unmodulated or unconditioned by any of the changing qualities of experience.

After what seems like an eternity, Jane opens her eyes and smiles at Clare. It is a look that is informed by the imperturbable peace of her essential being, and Clare feels its blessing power. She smiles back at Jane without saying anything, and as their eyes meet an ancient recognition stirs in them both.

And at that moment Mary wakes up.

*   *   *

Lying there in her bed in the middle of the night, Mary realises, 'It's only as Jane's mind that I could believe and feel that the people and the objects on the streets of London were something outside myself, separate from myself, made out of something other than myself, called matter. But I never really became Jane's mind. I never really experienced anything outside of myself. To experience the streets of London, I had to ignore, forget or overlook myself; in order to wake up as Jane's waking state I had to fall asleep.'

What is sleep from the point of view of Mary is waking from the point of view of Jane. What is sleep from the point of view of Jane is waking from the point of view of Mary. Mary recalls her study of the Bhagavad Gita: 'What is sleep for infinite awareness is waking for the finite mind; what is sleep for the finite mind is waking for infinite awareness.'

Mary realises, 'My mind is unlimited. My mind never really assumed the limits of Jane's finite mind. Jane was just a vehicle through which I was able to perceive London. But the knowing with which Jane knew her experience is *my* knowing! There is no real Jane, no real finite mind. Jane's finite mind is my infinite mind! The "I" of Jane is me, the only "I" there is. Jane thought she was practicing self-enquiry, tracing her experience back to its essential reality. But there is no Jane. What is, from Jane's point of view, a process of meditation or self-enquiry is, from my point of view, the process through which I simply unveil myself and stand naked and alone.'

Mary ponders her experience in this way but soon drifts off to sleep. In fact, she doesn't fall asleep; she simply allows her mind to assume the form of Jane again.

If Jane directs her attention outwards towards objects or inwards towards thoughts and feelings, the finite objects she encounters will appear as an inevitable counterpart to the finite mind with which they are known, for a finite mind can only know a finite object. However, in Jane's mind there is a doorway that does not lead in the direction of objects, be they external or internal. It is the doorway above which is written the name 'I' or 'I am'.

Jane recalls the image of the words *Know Thyself* carved above the entrance to the temple of Apollo in Delphi. If Jane's mind directs itself towards *itself*, that is, towards the knowledge 'I' or 'I am', it will, without necessarily realising it to begin with, be tracing its way back to its origin, that is, Mary's mind. If Jane asks herself the question, 'What is the nature of the knowing with which I know my experience?' or 'Who am I?' or 'Where do my thoughts and feelings come from?', her mind will direct its attention away from the objects it knows, towards the knowing with which it knows those objects, that is, towards the very nature of itself.

'I' is the name the mind gives to itself. 'I' is the essence of the mind. It is that part of Jane's mind that always remains the same, in all circumstances and states. In other words, 'I' is the hint in Jane's mind of its essential nature; it is the presence of Mary in Jane's mind. If Jane takes the thought 'I' and follows it, rather than directing her mind towards objective experience, her mind will trace itself back to Mary's mind.

Like Ariadne's golden thread, the 'I' thought is the path along which Jane's mind travels on its way back to Mary's mind, that is, on the way back to its true nature. The 'I' thought is the same path along which Mary has to travel in order to experience the streets of London as Jane. If Mary wants to experience anything other than herself, she has to overlook her own infinite nature, because her infinite nature can only know itself: infinite consciousness can only know infinite consciousness.

For there to be manifestation, consciousness must overlook, forget or ignore the knowing of its own infinite being. If Mary wants to know the streets of London, she has to turn away from the knowing of her own being and localise herself in and as Jane's body. It is through Jane's body that she can perceive the streets of London. The 'I' thought is the portal through which Mary, infinite consciousness, passes on its way to becoming Jane, the finite mind, and it is the same portal through which Jane passes in the opposite direction in order to experience the peace of her true nature.

The 'I' in Jane's mind is the hint of Mary. In fact, the 'I' of Jane *is* Mary! However, once Mary's infinite mind has assumed the form of Jane's finite mind, the knowledge 'I' or 'I am' gives Jane a clue as to her essential, irreducible nature. Likewise, the very 'I' that each of us now feels is the hint of infinite consciousness in each of our minds. It is the trace of God's infinite presence shining in the finite mind. If the finite mind explores the 'I' around whom its experience revolves, that exploration will take the finite mind on a directionless journey, a pathless path, in which it is gradually – or occasionally suddenly – relieved of its limitations and stands revealed as infinite consciousness.

If it is clear to us that the self that we essentially are belongs not to this body but to infinite consciousness, it doesn't imply that we stop experiencing the world and others from the perspective of the body. Consciousness continues to take the form of a body and appear as a finite mind. Mary continues to dream she is Jane, but now she is lucid dreaming. She sees and knows everything as herself. Likewise, once the finite mind has recognised that the knowing with which it knows its experience belongs to infinite consciousness, that recognition profoundly conditions the way we relate to experience. In fact, it is only infinite consciousness that recognises itself, because infinite consciousness is the only consciousness there is.

It is not Jane who recognises she is Mary. There is no Jane. It is Mary who recognises she is Mary. The finite mind cannot recognise infinite consciousness, because there is no independently existing finite mind in the first place. The moon's light cannot illuminate the sun because there is no real moonlight. The moon's light already belongs to the sun. There is just infinite consciousness, veiling itself in the form of the finite mind, and then unveiling itself. It is for this reason that Balyani said, 'I knew my Lord through my Lord.'

After this recognition we continue to experience objects, others and the world from the point of view of a separate subject of experience, but all are known, seen and felt to be our very own self. *Maya* remains but ignorance goes. As a result of this recognition there is a gradual purification or reorientation of the thoughts and feelings, and subsequent activities and relationships, that were previously based on the belief that 'I' – the knowing with which our experience is known, and out of which it is made – is temporary and finite.

In the Christian tradition this is known as the Transfiguration. It is the outshining of all experience in the light of pure knowing, or, in religious terms, in God's infinite being. In time this recognition is reflected and expressed in the absence of fear, lack or limitation and, as a result, the prevalence of love, compassion, justice, equity and humour.

CHAPTER 18

THE SEARCH FOR HAPPINESS

The body appears in the mind as a series of sensations and perceptions, and the mind is a vibration of awareness. As such, the body is not something solid made out of matter but a condensation or localisation of and in awareness.

In order to effect this condensation or localisation, the unlimited, space-like presence of awareness must contract or collapse within itself. This contraction is an *activity* of awareness, and its maintenance requires energy. The contraction of awareness into a finite mind exerts a tension on itself which is always seeking to be relieved, just as the compression of a rubber ball sets up an inevitable tension that is always trying to expand or relax into its original neutral condition. This contraction of awareness is felt as the experience of suffering, and the inexorable force toward the natural state of equilibrium is felt as the desire for freedom, peace and happiness.

Thus, the desire for happiness is simply the mind's desire to be divested of its limitations and returned to its inherently relaxed, peaceful condition of eternal, unlimited awareness. The search for enlightenment is simply a refinement of the desire for happiness. It is an indication that the search has become conscious rather than simply instinctive.

All human beings retain the memory of their own eternity within themselves, and it is for this reason that everyone without exception is motivated to seek freedom, peace, happiness and love above all else. Just as the screen is still visible in the image, and the echo of the Big Bang is said to be still discernible in the universe in the form of cosmic background radiation, so the original nature of awareness, which is peace, happiness and freedom itself, is discernible throughout all the mind's activities. The intuition of happiness is the echo of our true nature reverberating in the finite mind.

Every drop of water is a temporary name and form of the ocean. Although every drop is unique in terms of name and form, each one carries the same essential taste of the ocean. Likewise, every moment of experience is a temporary colouring of awareness, each moment unique in itself but carrying the same essential taste. The longing for happiness that lives in the hearts of all apparently separate selves is the desire to savour this one taste.

Awareness seems to overlook or forget itself in order to take the form of manifestation, but even in its apparent forgetting or veiling of itself it retains the memory of its own original nature. This memory, which expresses itself as the desire for happiness, is the pull of our original nature filtering through all forms of experience. It is an expression of the innate and inexorable force that exists in awareness itself to return again and again to its original, unconditioned, inherently peaceful and unconditionally fulfilled nature. It is for this reason that the desire for happiness eclipses all other desires. To find happiness is the ultimate purpose of human existence.

* * *

It is not in fact the human being that desires happiness. The human experience is a flow of mind – a spectrum of experience ranging from the subliminal states of the collective and personal 'unconscious' to the more clearly defined and sharply focused forms of the waking state. Each mind is a pulsating flux of energies through which, in which and as which awareness realises a segment of its infinite possibilities, and is thus a partial actualisation of itself. The individual mind is the agency by which awareness seems to become a separate subject of experience, from whose perspective it is able to know objective experience. As such, duality is the mechanism of creation.

However, this partial actualisation of awareness involves a trade-off: awareness must consent to limit itself in order to realise a segment of its infinite potential. In doing so, it allows its nameless, formless being to assume a name and a form. Through this consent, being appears as existence. In order to bring manifestation out of being and into existence, the infinite contracts into the finite. The tension that is created by this contraction is the longing for happiness.

From the point of view of the individual, happiness is something it desires for its own sake. Little does the individual realise that the desire for happiness is simply an equal and opposite force seeking to relax or dissolve the tensions that are inherent in the limiting of awareness from which it derives its apparent existence. The individual does nothing. It doesn't even have a status of its own. It is an activity, not an entity.

The entire existence of the apparently separate self or finite mind and the world that it perceives is a play in and of awareness. Awareness itself breathes the world into apparent existence at the expense of its own innate happiness. In doing so it seems to become so intimately merged with every aspect of its creation as to lose itself in it, and then it reclaims that happiness as its dissolves its self-assumed limitations and returns to itself. It is as though awareness breathes the separate self out of itself on the exhale, which is immediately followed by the natural impulse to draw in its breath.

The desire for happiness is the gravitational pull of our true nature on itself when it has lost itself in its own imagination. It is the pull from our true nature of perfectly free and inherently fulfilled awareness on the limitations of the finite mind. From the point of view of the individual, this pull is felt as desire or longing. However, that is said as a concession to the apparent individual from whose imaginary point of view it seems to have an independent existence of its own. In reality, no such separate self ever comes into existence, and therefore there is no question of such a self returning to its true nature.

It is because the screen is present everywhere in the movie that it can never appear as a specific object *in* the movie and seems, from the point of view of a character in the movie, to be lost in it and therefore missing. Likewise, it is because awareness pervades *all* experience so intimately and homogeneously that it can never become a particular object *of* experience, and thus, from the point of view of the apparently separate self or finite mind from whose perspective experience is known, seems to be missing.

Awareness only seems to be missing because it is so completely present in all aspects of its creation that it cannot be distinguished from it. From its own point of view, there is nothing in awareness other than awareness itself, and therefore there is no question of ever experiencing its

THE NATURE OF CONSCIOUSNESS

own absence. Awareness seems to be nowhere because it is everywhere. It seems to be nothing because it is everything.

Once awareness has apparently veiled itself by freely assuming the limitations of the body, it seems to cut itself off from the knowing of its own innate peace and happiness. That is why the apparently separate self feels a wound at the heart of its being, a sense that something is missing or has been lost. It is for this reason that John Bunyan said that God enters the soul through a wound. This wound initiates the search for happiness which is the defining characteristic of the separate self.

The separate self is not an entity; it is the *activity* of this search. The separate self does not *feel* this wound; it *is* this wound. It is not the self that moves towards God; it is God that attracts the self. The movement of the self towards happiness is called desire; the pull on the self from happiness is called grace. As the sixteenth-century Italian monk realised, 'Lord, Thou art the love with which I love Thee.'

When the individual realises that its entire experience is always tending to return to its natural condition, it realises that it does nothing. Its desire for happiness is simply its response to God's invitation to return, the grace that is the inevitable pull of awareness acting on itself whenever it seems to have forgotten or overlooked its true nature.

The repeated impulse to return to its natural condition may be initiated in the life of the separate self or finite mind by an object, person or teaching whose sole purpose is to effect the dissolution of the finite mind into its infinite source, and may take place on many different timescales: at the end of every thought or perception, at the end of every day and at the end of every lifetime.

*   *   *

Long before I was able to articulate this clearly, I had my first intuition of it, as indeed most people do, although this intuition is often overlooked for lack of proper guidance. A couple of years after the crisis at school described in Chapter 5, I found myself living on the edge of Bodmin Moor, in southwest England, apprenticed to one of the founding fathers of the studio pottery movement in the latter half of the twentieth century. Michael Cardew was an old Zen master: cantankerous and irascible on the one hand, warm-hearted and kind on the other. Coupled with a ruthless and penetrating intellect, this made him a disarming and formidable character, which for a young, idealistic man seeking meaning outside the parameters for which his education had thus far prepared him, was an intoxicating and irresistible invitation.

However, it came at a price: life on the edge of Bodmin Moor was solitary and spartan – perhaps intolerably so, were it not for the solace of a friend. But I had a friend. Every Friday evening after dinner I would walk a mile or so up the lane into the village, and call her from the phone box which stood on a small triangle of grass at a junction in the road that led across the moor. That Friday evening, like so many before, I walked up to the phone box seeking refuge. Everything I needed to know was contained in her greeting. I didn't hear anything after that.

Even as I walked down the hill, I intuited that the dilemma that had appeared to me at school a few years earlier had now taken on a new dimension and was about to intensify. The question as to what aspect of the mind's knowledge could be relied upon was no longer simply of interest; it had gripped me. This was no longer just about knowledge; it was about happiness. I felt it as a burning in my body, before it was rationalised in my mind. It would be many years before I realised that the search for understanding in my study of the Vedantic teaching, for beauty in my studio as an artist and for happiness and love in intimate relationships was the same quest.

That burning initiated a profound investigation into the nature of happiness, its source and the means by which it might be attained. If someone or something can at one time be a source of happiness and at another a source of suffering, in whom or in what can one reliably invest the desire to be happy? Without realising it, in that brief phone call, I had been given the greatest gift one can ever receive: the intense desire to find out the nature of lasting, unconditional happiness, and its source.

In theory, only one such experience should be required to make it clear that the cause of the heart's wound is not the absence of any object, state or relationship but rather the forgetting, ignoring or overlooking of our essential nature of ever-present and unlimited awareness, whose nature is peace and happiness itself. But in practice most of us need many such initiations in the form of failed relationships, misfortune, disillusionment or disappointment in order to realise this.

*　　*　　*

Just as a spider spins a web out of herself and then lives as a creature *within* that web, from which she now has to extricate herself, so consciousness imagines the world within itself and then identifies itself as one of the bodies in that world, from whose perspective it now seems to know it. Consciousness seems to become an inside self made of mind living in an outside world made of matter.

Before the spider spins her web, it lies in potential inside her. The moment she spins her web, she becomes a spider that lives *in* that web, which now seems to be outside and distinct from herself. The spider has been reduced to a fragment, and the web to which she has given birth now seems to be her host. The web and the spider have changed places.

In the same way, the world lies in potential in consciousness. Consciousness generates the world within itself and then, forgetting its true nature, enters its own imagination in the form of a separate self from whose perspective that world may be known. So deep is this amnesia that the separate subject now considers that the world made of matter in which it seems to live is primary and, as a result, that its own essential nature – consciousness – is a by-product of it.

However, the spider that *spins out* the web and the spider that *lives in* the web is the same in both cases. Likewise, the infinite 'I' of consciousness that generates the world within itself and the personal 'I' that seems to live in that world are the same 'I'. This is why Ramana Maharshi said, 'When the "I" is divested of the "I", only "I" remains.'

Having freely given up the knowing of its own eternal, infinite being and assumed the form and therefore the limits of the body, consciousness seems to have become its own prisoner. The prison in which it has incarnated itself is the body, and by doing so it seems to have acquired its limits and destiny.

Prior to this apparent division of itself through the ignoring of its own infinite reality, there is just infinite consciousness being, knowing and loving itself alone. Even *during* this apparent forgetting there is still only infinite consciousness. It is this infinite consciousness that takes the shape of thoughts, images and feelings on the 'inside' and sense perceptions on the 'outside', without ever ceasing to be or know itself alone. There is no other substance present in experience.

Why would consciousness freely do such a thing? We cannot give a reason. Any reason would itself be part of manifestation and, as such, part of the objective world for which we were seeking a cause. At best we can say that it is simply an overflowing of itself into manifestation, a sacrifice of its own inherent peace and freedom, an impulse of love in which pure consciousness or God's infinite being pours itself out into form for no reason, and then, finding itself imprisoned within its own creation, begins the return journey. As Hafiz says, 'It is an impulse of love for the sake of beauty.'

All apparently separate selves feel that they have free will and that this freedom is theirs by birthright, and for good reason. In the hearts of all apparently separate selves lives the memory of our eternity, the longing for freedom, happiness, peace or love, and it is impossible for that flame to be completely extinguished. The free will that each of us feels is an echo of the freedom of infinite consciousness, the freedom of God's infinite being. The exercise of that free will in the pursuit of happiness, peace or love is an impulse that cannot be satisfied by anything but the absolute truth and unconditional love.

*   *   *

By acquiring the limits of the body, consciousness appears to become a fragment and, as such, feels cut off from the whole, incomplete, lacking and alone. As a result, this consciousness-in-the-body entity – the ego or separate self – is in a perpetual state of desire, always seeking to relieve the sense of lack, incompletion and loneliness through the acquisition of objects, substances, activities, states of mind and relationships.

By seeming to share the destiny of the body, consciousness appears to become a temporary entity, subject to birth, change, ageing and death. It is for this reason that the consciousness-in-the-body entity lives with a deep fear of disappearance and death and is almost perpetually trying to allay this fear through emotional defence and resistance.

Thus desire and fear, or seeking and resistance, are the two essential activities around which the ego, separate self or consciousness-in-the-body entity revolves. In fact, the ego is not an entity with its own independent existence; it is the activity of desire and fear. Most people's lives are, without their realising it, dominated almost entirely by these two existential feelings, which lie for the most part unnoticed below the threshold of the waking-state mind, subliminally influencing most of their thoughts and emotions, and the subsequent activities and relationships that proceed from them.

In fact, most people's lives are spent avoiding ever having to fully face the discomfort of this existential lack and fear; it is an almost full-time activity that engages people with varying degrees of intensity in a variety of activities, substances and relationships. These strategies of avoidance work to a greater or lesser extent, although even in the most successful lives this existential lack and fear regularly percolate into everyday experience from the unseen depths of the mind, disturbing us with irrational thoughts and unwelcome feelings that are subsequently lived out in our activities and relationships.

To live a life based on the assumption of such an ego or self is to live a life of ignorance – or, in the Christian tradition, sin – a life in which the reality of experience is ignored or denied. A life so lived generates, perpetuates and communicates the ignorance at its core; hence the current state of our world culture, which is almost exclusively dominated by the illusion of separation. The mind/matter divide at the heart of this illusion is the hallmark of materialism and the foundation upon which all conflict and unhappiness are based.

All that the mind needs to do to know its own reality is to cease being exclusively fascinated by the objective elements of its experience – thoughts, feelings, sensations and perceptions – and ask itself instead about the nature of the knowing with which it knows that experience. To find the answer to this question, the mind must turn its knowing or attention away from the objective knowledge that it knows and redirect it towards itself, that is, towards the very knowing with which it knows that knowledge.

When this knowing gives its attention to itself rather than to any finite object or state, it doesn't find any limitations there. It doesn't find a finite mind, a limited consciousness. It finds its own nature: original mind. Even to say it 'finds original mind' is a concession to conventional language, suggesting that a subject finds or knows an object. This finding is more like a recognition, a divesting of the finite mind of its self-assumed limitations, leaving its original nature – pure consciousness – revealed.

Nothing new is found in this recognition; layers of obscuration only fall away. It is referred to as a recognition because it is not something new that is known; it is rather something that was forgotten that has been remembered. It is a revelation. The word 'revelation' comes from the Latin *revelare*, meaning 'to lay bare'. It is a laying bare of that which was previously obscured by finite thought and perception. At that timeless moment

– timeless because, as the limitations of mind fall away, so time itself dissolves – the apparently finite mind loses its finiteness and thus ceases to be mind, as such. It is revealed as pure consciousness – empty, transparent, dimensionless, objectless, limitless, non-dual, self-aware being.

In all people, under all circumstances and in all situations, the memory of our eternal nature – original mind or pure consciousness – remains alive, however obscured it may seem to be at times. When it seems to be obscured, this memory expresses itself as a longing for truth, happiness, peace, love or beauty. These desires are all facets of a single desire: the mind's desire to be divested of its self-assumed limitations.

* * *

I recently spent an afternoon walking with my friend Bernardo Kastrup through the streets of Amsterdam, experiencing, as he put it, an aspect of the city that I would not normally encounter on my circuit of non-dual meetings. We walked through a funfair in which groups of teenagers were bungee jumping in capsules; sat for some time in a church in which a mass was taking place; stopped for a drink outside a café; walked through the red light district; and visited one of Amsterdam's notorious 'head shops' before returning to our hotel.

As we walked, I could not help but notice that nearly everyone we encountered seemed to be seeking, in one way or another, to relieve the discomfort of the existential lack and fear that lie as a wound in the hearts of almost all people. As the teenagers plummeted in free-fall from the height of their ascent, they felt the fear of death from the safety of their capsule, and the immense relief from that feeling when they finally came to rest. In this self-imposed initiation rite, they tasted and survived the terror of death and, as a result, felt for a few moments the joy of their own unconditioned existence, before the conditioned mind reasserted its strategies of denial and avoidance and eclipsed the peace and fulfilment that lie at its source. In that brush with death, the teenagers' existential fear was exposed and fully felt, and in surviving the ordeal they briefly tasted that element in themselves, their essential being, that lies deeper than the ego. The sole purpose of the jump, the brush with death and the exposure of the fear was to artificially induce the taste of their own eternity.

In the enactment of the mass, people were similarly seeking to be released from the limitations of the ego. By surrendering everything in themselves to a higher power, they were, as it were, emptying themselves of the burden of the ego with its train of desire, fear, neurosis, conflict, confusion, doubt and agitation, enabling themselves to savour their essential being, free of conditioning, in all its innocence, purity and peace. Rather than dissolving the ego in its source, such devotional practices expand it beyond its customary limitations, releasing it from its self-contracted state and commending it to God's infinite being, in which it finds rest and peace. In the words of Isaiah, 'Thou wilt keep in perfect peace whose mind is stayed on Thee.'

On the cobbled street outside the café, the first few sips of cool beer relax the activity of the mind with which the ego defines and perpetuates itself. As the activity of the mind relaxes, it expands and begins to sink backwards into its source of pure awareness. Even a few steps in this direction are enough to relieve the mind of a degree of its agitation, and as it continues to expand with further sips, it is progressively relieved of the contraction from which the ego derives its identity, affording the mind the fragrance if not the full taste of its own essential nature of peace and freedom. As the person looks around at the activity on the streets of Amsterdam, he does so now as a spectator and not a participant. For a few minutes the relaxation of his mind allows him to stand as the witness of his experience, no longer its accomplice, and as a result he experiences the innate peace and fulfilment of his true nature.

In gazing at an almost naked young woman from the distance of a metre and separated only by a pane of glass, the sense of lack, insufficiency and inadequacy that live at the core of the separate self or ego is exposed and further heightened by the promise of its immediate and gratuitous fulfilment. The subsequent consummation of his desire allows the man to mimic the motions of intimacy without ever having to pay the real price of openness and vulnerability, and at the same time puts a temporary end to the discomfort inherent in his longing, the degree of relief experienced being commensurate with the intensity of the desire evoked. This exposure and fleeting dissolution of the sense of lack that lives at the core of the ego divests the mind temporarily of its limitations, allowing it to plunge, as it were, into its source and taste its essentially unlimited and unconditioned nature, which the man experiences as peace and happiness.

In the head shop, a vast array of mind-altering substances are on offer, all of which promise to relax and expand the mind beyond the prison in which it has located itself and, as a result, to give it a taste of its original, unconditioned and inherently free nature. As the mind relaxes and expands, it travels backwards or inwards through the broader medium of its own field, visiting experiences that are not available to it in the waking state. These experiences give the mind a hint of its own limitless possibilities, of which its waking-state experience is only the narrowest realisation. As the narrow focus of the waking state is relaxed, the distinction between the objects and selves that it experiences becomes less and less clearly defined, and the shared field in which they arise and of which they are but modulations becomes increasingly obvious. The underlying unity of all objects and selves begins to become self-evident. However, attracted by the relative freedom experienced while exploring the broader medium of its own potential, and yet rarely, if ever, glimpsing the absolute freedom of its own inherent nature, the mind becomes addicted to such states and returns to them again and again, seduced by their promise of freedom and simultaneously bound by their limitations.

In all these cases, the person in question wrongly attributes the peace, happiness and freedom briefly experienced to the acquisition of the object, activity, substance, state of mind or relationship and, as a result, when the underlying suffering resurfaces again in between the normal activities of the outward-facing or object-seeking mind, he simply returns to the same objective experience, hoping thereby to experience the same relief, in an ever-deepening cycle of longing, addiction and despair, each time requiring a slightly stronger dose of the object to achieve the desired result.

Unlike the Tantric practitioner, who allows her desire to be aroused but surfs it inwards to fulfilment in its source rather than pursuing it outwards towards the object, substance or state, the seeking mind becomes progressively addicted to the objective experience that seemed to precipitate the brief experiences of peace and happiness.

The finite mind is always seeking to dissolve or expand itself, to divest itself of its limitations and return to its original, unconditioned nature, and thus to taste the peace, happiness and freedom that reside there simply waiting to be recognised.

The essential, irreducible essence of the mind – the absolute truth of experience that shines in each of us as the experience of being aware, the knowledge 'I am' or the feeling of love, and which is known variously as 'I', consciousness, awareness or God's infinite being – is that aspect of mind that cannot be removed from it and is common to all beings. Indeed, it is common to all existence.

It is equally available to all people, at all times and under all circumstances, and it is the foundation of peace within individuals, families, communities and nations. As such, it must be the foundation of civilisation. To found a civilisation upon any other knowledge is to build a house on the shifting sands of local, temporal belief, and this can never be the basis for true community, tolerance and harmony.

All that is required is for the mind to notice that its own essential existence is shared with the existence of all beings and things, and to live the implications of that recognition in all realms of life.

Under Rupert's gentle but decisive guidance, you have just explored the underlying nature of reality through the primary – yet most neglected – avenue of knowledge available to us: *introspection*. Rupert's mastery of introspection, and his ability to take us along with him as he explores the foundations of Self and World, reveal what our cultural indoctrination has laboriously kept hidden from us: that there is, in fact, no difference between the two. Self and World are one, a conclusion as contrary to our mainstream cultural narrative as it is self-evident upon lucid introspection.

How can there be such dissonance between the basic tenet of our culture and direct introspective experience? Even if this book has succeeded in helping you truly *understand* that the World is an excitation of the Self – in Rupert's words, 'a movement of mind' – no more distinct from the latter than ripples are distinct from water, the power of the mainstream cultural narrative may still instigate a lingering discomfort. 'Is it plausible that our entire culture could have gotten it so wrong?' you might ask yourself. In this brief Afterword, I will attempt to show you that, because of an imbalance in our culture's approach to knowledge, not only is this plausible but it is to be expected.

You see, we can acquire knowledge through three distinct avenues: empirical observation, rational thought and introspection. Empirical observation consists in the subset of our experiences associated with the five senses. As such, if we define the World as encompassing everything we can see, hear, touch, taste and smell, then empirical observation consists in knowing the World directly. Notice that, defined in this way, the World is simply a set of experiences qualitatively equivalent to, for instance, personal imagination. Yet it differs from personal imagination in that it is collective rather than idiosyncratic: after all, we all seem to share the same World. Empirical observation of this collective World is thus an avenue of knowledge orthogonal to the imagination, as history painfully illustrates. Aristotle, for instance, *imagined* that heavier objects fell to the ground faster than lighter objects,[1] an idea that persisted for almost two millennia. Only when Galileo decided to *empirically observe* whether that were really the case – by famously dropping two canon balls of different weights from the leaning tower of Pisa – did we realise that the World is, in fact, different from what Aristotle had imagined it to be. Thanks to empirical observations such as Galileo's, we have now been able to know the World well enough to put a man on the moon and robots on Mars, and even land a probe on a comet.

By empirically observing the World we can discern its patterns and regularities during observation. But to *infer* how the World behaved before observation and *predict* how it will behave in the future, we need to *model* those patterns and regularities in the form we have come to call the 'laws of nature'. And here is where the second avenue of knowledge comes in: *rational thought* enables us to deduce unobserved – and even unobserv*able* – aspects of the World from observed ones. It allows us to connect the dots and extrapolate the boundaries of our knowledge beyond what can be directly apprehended through the five senses. It is rational thought that, for instance, enables engineers to know which building design will stand firmly and which phone design will communicate reliably *without* having to try out every possible variation. It is also rational thought that enables us to infer explanations such as the Big Bang and hominid evolution, even though we cannot empirically observe either. Rational thought provides the template along which both explanatory and predictive models are woven.

The third and final avenue of knowledge is, of course, introspection. By introspecting, we turn our attention from the World to the *knower* of the World and the *process of knowing*. We ask: Who or what is it that knows? How does it know what it knows? As Rupert says, 'What is it that knows or is aware of my experience? What is the nature of the knowing with which all knowledge and experience are known?' Knowledge only has meaning insofar as these questions are answered. After all, as a state of the knower and the outcome of the process of knowing, knowledge is secondary to both. Everything we believe we know through the other two avenues – empirical observation and rational thought – is thus ultimately conditioned by introspection. Whatever information we derive from observation and thought only has meaning insofar as we understand the nature of the knower and how it knows. Without such understanding, the natural patterns discernible through observation and thought are akin to ripples without water, choreographies without dancers, spin without tops. They delineate an empty mould whose actual substance can only be filled through introspection.

And here is where the problem lies. Introspection requires an intimate engagement with the *subject* of experience, as opposed to its objects. But science – whose values and methods have informed our mainstream cultural narrative for the past two centuries or so – must stand clear of subjectivity. As Rupert explains, 'In its search for the absolute truth, science rejects subjective experience on the grounds that it is personal and therefore cannot be validated by anyone other than the person having the experience.' This is entirely appropriate insofar as one chooses – as science does – empirical observation and rational thought as one's *sole* avenues of knowledge. After all, as discussed above, the World is defined as the shared subset of our experiences, as opposed to the figments of one's personal imagination. So to properly assess the World, science must indeed set aside idiosyncratic reports and focus on experiences consistently shared by multiple individuals.

But whilst internally consistent, the scientific method is *incomplete* in that it disregards true introspection. As such, it is ill-equipped to answer any of the fundamental questions about the nature of the knower and the process of knowing. Correctly understood, science merely models the patterns and regularities of the World without providing any insight into its underlying nature. It doesn't tell us what the World *is*, only how it *behaves*. It characterises the choreography without saying anything about the dancer. It predicts the ripples without saying anything about the water. It describes the spin without saying anything about the top. In Rupert's words, 'Most people believe that science is gradually inching its way towards an understanding of the fundamental reality of the universe. However, until consciousness itself becomes the focus of scientific interest, researchers will still be seeking the fundamental reality of the universe in a thousand years' time.'

Be that as it may, it seems difficult for most scientists to acknowledge the inherent limitations of their method. As a former professional scientist myself, I base this assertion on my own personal experience. Scientists have a natural tendency to believe that they are unveiling what the World *is*, not just how it *behaves*. Believing otherwise would detract from much of the romantic allure that brought scientists to their profession in the first place. Moreover, it is admittedly difficult – at a psychological level – to do science without at least a working hypothesis for interpreting the patterns and regularities discerned through experiments. Stanford physicist Andrei Linde, renowned for his theories of cosmological inflation, explained it best:

> Let us remember that our knowledge of the world begins not with matter but with perceptions…. Later we find out that our perceptions obey some laws, *which can be most conveniently formulated if we assume that there is some underlying reality beyond our perceptions*. This model of a material world obeying laws of physics is so successful that soon we forget about our starting point and say that matter is the only reality, and perceptions are only helpful for its description. This assumption is almost as natural (and maybe as false) as our previous assumption that space is only a mathematical tool for the description of matter.[2]

So what was originally a mere working model to facilitate the interpretation of scientific observations has now hardened into the dogma of a material world outside mind. This hasty jump was driven by the psychological need to fill a vacuum: scientists couldn't operate without a way to think about the World in terms of its underlying reality. While constructing abstract characterisations of the choreography, they needed a way to visualise the dancer. And then, because proper introspection had never been part of their professional skillset, *they simply took the World at face value*. To this day we pay the price for such a lazy blunder. Indeed, if scientists knew what you now know after having read this book, they would surely have thought things through a little more carefully.

In and by itself, the blunder of associating science with materialism would probably have been of limited consequence. But in conjunction with a second blunder, it has had the devastating effect of causing our culture to dismiss all legitimate paths to true insight. This second blunder is our culture's elevation of science – an incomplete method – to the position of ultimate arbiter of truth, as opposed to a pragmatic approach for producing technology and informing philosophy.

You see, because we tend to conflate what works with what is true – an error easily seen when considering theories that work in practice yet aren't actually true, such as Newtonian mechanics and Fourier optics – we mistake the technological success of science for evidence that it provides insight into the underlying nature of reality. This is akin to believing that a five-year-old kid who plays computer games very well understands the underlying nature of computer hardware and software. Mistaking effectiveness for understanding, our culture proclaims that the scientific method is the best way to figure out what the World *is*, not just how it *behaves*. Consequently, we now have a one-eyed pilot overloaded with the heavy baggage of materialism trying to guide our flight towards truth. The baggage is so heavy one must wonder whether we can even leave the ground, let alone find our way.

Science's emphasis on empirical observation and rational thought, at the cost of true introspection, is the imbalance in our culture's approach to knowledge that I alluded to in the beginning of this Afterword. But as the root cause of the problem, the imbalance is also the obvious place to apply a fix. Indeed, restoring some lucid introspection to science can initiate a far-reaching domino effect by revealing how *both empirical observation and rational thought themselves indicate that Self and World are one.* Allow me to elaborate on this perhaps surprising claim.

In a famous 1960 paper titled 'The Unreasonable Effectiveness of Mathematics in the Natural Sciences',[3] renowned physicist Eugene Wigner discussed what he described as 'the miracle of the appropriateness of the language of mathematics for the formulation of the laws of physics'. Indeed, mathematical methods and results envisioned purely in abstraction have, again and again, turned out to precisely describe concrete aspects of the World. For instance, non-Euclidean geometries, whose axioms assume space to be curved, were developed at a time – the early nineteenth century – when everyone 'knew' that space was flat. So these non-Euclidean geometries, although mathematically rigorous, were considered fictions, models of imagined things whose validity resided only in the minds of mathematicians. Within only a few decades, however, Einstein found out that space is, in fact, curved, a fact confirmed through empirical observations. Non-Euclidean geometries then turned out to describe the World itself with uncanny precision and accuracy.[4] Their validity thus somehow extends far beyond the minds of mathematicians.

Why and how entirely abstract creations of rational thought – based solely on axiomatic intuitions – turn out to describe the structure and dynamics of the World at large remains a profound mystery to this day, at least under the materialist paradigm.[5] In Wigner's words, 'It is difficult to avoid the impression that a miracle confronts us here, quite comparable in its striking nature to the miracle that the human mind can string a thousand arguments together without getting itself into contradictions.' The 'miracle' (Wigner uses this word twelve times in his paper) is perhaps most pronounced in quantum mechanics, where – as reflected in the famous admonition 'Shut up and calculate!' – *only the mathematics* is clearly understood, not the actual World it so accurately models.

It is tempting to try to pull ourselves up by our own bootstraps and simply proclaim that the axioms of logic and mathematics should be applicable to the World at large. But lest we fall into the fallacy of circular reasoning, we cannot logically argue for the validity of logic beyond our minds, so the World might as well be absurd.[6] By the same token, under the postulate that Self and World are distinct there is just no reason to think that the World should comply with abstract mathematical truths devised in mentation. Why should it? Yet we know empirically that it does, which baffles – as it should – the materialist mind-set.

Under the non-dual view expressed in this book, on the other hand, the correspondence between the intuitive foundations of *rational thought* – as reflected in the axioms of logic and mathematics – and the way the World works is perfectly natural. Indeed, it couldn't be any different. You see, that we take the basic tenets of logic and mathematics to be self-evident truths betrays their *archetypal nature* in the Jungian sense: they are irreducible psychological templates according to which thought unfolds.[7] As a matter of fact, Marie-Louise von Franz went as far as to argue that the natural numbers themselves are archetypal.[8] Then – and here is the key point – *the fact that these archetypes extend into the World clearly indicates that the World itself is mental and continuous with the Self.* Even modest introspection suffices to see this. If there is no separation between mind and the objects of perception, of course these objects should comport themselves in a way consistent with the psychological archetypes of mind. Perceptual objects should be an expression of archetypal patterns in the same way that thoughts and emotions are, so the World should be consistent – as it is – with our logic and mathematics. The apparent eeriness of Wigner's 'miracle' thus melts under the non-dual view articulated here by Rupert like butter under the sun. The alleged mystery is revealed by introspection to be a mere artefact of the confused materialist paradigm. That our culture at large still hasn't taken the hint is a reflection of the appalling state of our collective ability to introspect.

Not only the empirical validity of rational thought suggests the unity of Self and World; empirical observations also point to this unity, even more directly. Indeed, a key implication of the posited separation between Self and World is that the properties of the World should not depend on observation; that is, a perceptual object should have whatever properties it has – weight, size, shape, colour, and so on – regardless of whether or how it appears on the screen of perception. But this has statistical implications that can be directly tested.[9] On this basis, Gröblacher and others have shown empirically that the properties of the world *do* depend on observation.[10] To reconcile their results with materialism would require a tortuous redefinition of what we call 'objectivity'. And since our culture has come to associate objectivity with reality itself, the science press felt compelled to report on this study by pronouncing, 'Quantum physics says goodbye to reality'.[11]

Other statistical implications of the posited separation between Self and World[12] have also been experimentally tested, empirically demonstrating that the properties of physical systems *do not even exist* prior to being observed.[13] Commenting on these results, renowned physicist Anton Zeilinger has been quoted as saying that 'there is no sense in assuming that what we do not measure about a system has [an independent] reality'.[14] Finally, Ma and others have again shown, in 2013, that no naïvely objective view of the World can be true, in view of empirical observations.[15]

Critics have deeply scrutinised the studies cited above to find possible loopholes, implausible as they may be. In an effort to address and close these potential loopholes, Dutch researchers performed an even more tightly controlled test, which once again confirmed the earlier conclusions.[16] This latter effort was considered by *Nature News* magazine the 'toughest test yet'.[17]

Another implication of the posited separation between Self and World is that our choices can only influence the World – through our bodily actions – in the present. They allegedly cannot affect the past. As such, the part of our story that corresponds to the past must be unchangeable. Contrast this to the sphere of mind, wherein we can change the whole of an imagined story at any moment. In mind, the *entire* narrative is always acquiescent to choice and amenable to revision. Now, as it turns out, Kim and others have shown empirically that observation not only determines the physical properties observed at present, *but also retroactively changes their history accordingly.*[18] This suggests that the past is created at every instant so as to be consistent with the present, which is reminiscent of the notion that the World is a malleable mental narrative.

Already back in 2005, renowned Johns Hopkins physicist and astronomer Richard Conn Henry had seen enough: he penned an essay for *Nature* magazine wherein he claimed that 'the universe is entirely mental'.[19] As we have seen, empirical observations since then overwhelmingly corroborate his case. Yet many physicists refuse to acknowledge it. They postulate all kinds of unprovable invisible entities and try to develop tortuous mathematical acrobatics to find a way around the evidence. In the words of Conn Henry: 'There have been serious [theoretical] attempts to preserve a material world – but they produce no new physics, and serve only to preserve an illusion.'[20] The illusion he was referring to was, of course, that of a World outside mind; a World separate from the Self. The inability of many physicists to acknowledge what observations are telling us reflects, once again, a failure of introspection. The identity of Self and World is indeed hard to accept if one cannot look within to see it, getting stuck instead at face-value appearances.

In conclusion, when informed by even modest introspection, both rational thought and empirical observations indicate, in and of themselves, a unity between Self and World. All three avenues of knowledge thus point in the same direction. It is the lack of introspection in our culture's way of relating to reality that prevents us from seeing this. My intention in this Afterword has been to highlight the insidious effects of this lack, so as to help you recognise the critical importance of the present book. By masterfully restoring introspection to the cultural dialogue, Rupert addresses the root cause of our predicament. And by having read this book, you now find yourself in a privileged position to help tip the balance of things in favour of truth. Goodness knows we need it.

Bernardo Kastrup
September 2016

NOTES

1. Aristotle, *Physics*.

2. Linde, A., 'Universe, Life, Consciousness', a paper delivered at the Physics and Cosmology Group of the Science and Spiritual Quest program of the Center for Theology and the Natural Sciences, Berkeley, California, 1998 (emphasis added).

3. Wigner, E., 'The Unreasonable Effectiveness of Mathematics in the Natural Sciences', *Communications in Pure and Applied Mathematics* (1960).

4. See, for instance: Wilson, E. and Lewis, G., 'The Space-Time Manifold of Relativity. The Non-Euclidean Geometry of Mechanics and Electromagnetics', *Proceedings of the American Academy of Arts and Sciences* (1912).

5. In 2015, PBS released a documentary film in its *NOVA* series, titled 'The Great Math Mystery: Is math invented by humans, or is it the language of the universe?', that showed many surprising ways in which mathematical thought corresponds to the World.

6. For a more rigorous argument, see: Albert, H., *Treatise on Critical Reason* (Princeton University Press, 1985).

7. An analogy may help explain what psychological archetypes are: If mind were a vibrating surface, then the archetypes would be akin to the constraints that determine the natural modes of vibration of the surface. For more elaboration, see: Jung, C., *The Archetypes and the Collective Unconscious* (Routledge, 1991).

8. Franz, M.-L. von, *Number and Time* (Northwestern University Press, 1974).

9. Leggett, A., 'Nonlocal hidden-variable theories and quantum mechanics: An incompatibility theorem', *Foundations of Physics* (2003).

10. Gröblacher, S. et al., 'An experimental test of non-local realism', *Nature* (2007).

11. Cartwright, J., 'Quantum physics says goodbye to reality', *IOP Physics World* (2007).

12. Bell, J., 'On the Einstein Podolsky Rosen paradox', *Physics* (1964).

13. Lapkiewicz, R. et al., 'Experimental non-classicality of an indivisible quantum system', *Nature* (2011); as well as Manning, A. G. et al., 'Wheeler's delayed-choice gedanken experiment with a single atom', *Nature Physics* (2015).

14. Ananthaswamy, A., 'Quantum magic trick shows reality is what you make it', *New Scientist* (2011).

15. Ma, X.-S. et al., 'Quantum erasure with causally disconnected choice', *Proceedings of the National Academy of Sciences of the USA* (2013).

16. Hensen, B. et al., 'Experimental loophole-free violation of a Bell inequality using entangled electron spins separated by 1.3 km', *arXiv:1508.05949 [quant-ph]* (2015).

17. Merali, Z., 'Quantum "spookiness" passes toughest test yet', *Nature News* (2015).

18. Kim, Y.-H. et al., 'A delayed choice quantum eraser', *Physical Review Letters* (2000).

19. Conn Henry, R., 'The mental universe', *Nature* (2005).

20. Ibid.

PUBLICATIONS BY RUPERT SPIRA

The Transparency of Things – Contemplating the Nature of Experience
Non-Duality Press 2008
Sahaja Publications & New Harbinger Publications 2016

Presence, Volume I – The Art of Peace and Happiness
Non-Duality Press 2011
Sahaja Publications & New Harbinger Publications 2016

Presence, Volume II – The Intimacy of All Experience
Non-Duality Press 2011
Sahaja Publications & New Harbinger Publications 2016

The Ashes of Love – Sayings on the Essence of Non-Duality
Non-Duality Press 2013
Sahaja Publications 2016

The Light of Pure Knowing – Thirty Meditations on the Essence of Non-Duality
Sahaja Publications 2014

Transparent Body, Luminous World – The Tantric Yoga of Sensation and Perception
Sahaja Publications 2016

www.rupertspira.com